AF581799

High-Frequency Vacuum Electron Devices

High-Frequency Vacuum Electron Devices

Editors

Jinjun Feng
Yubin Gong
Chaohai Du
Adrian Cross

MDPI • Basel • Beijing • Wuhan • Barcelona • Belgrade • Manchester • Tokyo • Cluj • Tianjin

Editors

Jinjun Feng
National Key Labotatory of Science and Technology on Vacuum Electronics
Beijing Vacuum Electronics Research Institute
Beijing
China

Yubin Gong
School of Electronic Science and Engineering
University of Electronic Science and Technology of China
Chengdu
China

Chaohai Du
School of Electronics
Peking University
Beijing
China

Adrian Cross
Department of Physics and SUPA
University of Strathclyde
Glasgow
United Kingdom

Editorial Office
MDPI
St. Alban-Anlage 66
4052 Basel, Switzerland

This is a reprint of articles from the Special Issue published online in the open access journal *Electronics* (ISSN 2079-9292) (available at: www.mdpi.com/journal/electronics/special_issues/high_frequency_vacuum_electron_devices).

For citation purposes, cite each article independently as indicated on the article page online and as indicated below:

LastName, A.A.; LastName, B.B.; LastName, C.C. Article Title. *Journal Name* **Year**, *Volume Number*, Page Range.

ISBN 978-3-0365-5448-8 (Hbk)
ISBN 978-3-0365-5447-1 (PDF)

Contents

About the Editors . . . . **vii**

Preface to "High-Frequency Vacuum Electron Devices" . . . . **ix**

Jinjun Feng, Yubin Gong, Chaohai Du and Adrian Cross
High-Frequency Vacuum Electron Devices
Reprinted from: *Electronics* **2022**, *11*, 817, doi:10.3390/electronics11050817 . . . . **1**

Mengshi Ma, Qixiang Zhao, Kunshan Mo, Shuquan Zheng, Lin Peng and You Lv et al.
Design of a Dual-Mode Input Structure for K/Ka-Band Gyrotron TWT
Reprinted from: *Electronics* **2022**, *11*, 432, doi:10.3390/electronics11030432 . . . . **5**

Yidan He, Zhiwei Li, Shuyu Mao, Fangyuan Zhan and Xianlong Wei
A Vacuum Transistor Based on Field-Assisted Thermionic Emission from a Multiwalled Carbon Nanotube
Reprinted from: *Electronics* **2022**, *11*, 399, doi:10.3390/electronics11030399 . . . . **17**

Yuxin Wang, Yang Dong, Xiangbao Zhu, Jingyu Guo, Duo Xu and Shaomeng Wang et al.
Multiple Dielectric-Supported Ridge-Loaded Rhombus-Shaped Wideband Meander-Line Slow-Wave Structure for a V-Band TWT
Reprinted from: *Electronics* **2022**, *11*, 405, doi:10.3390/electronics11030405 . . . . **27**

Ruichao Yang, Lingna Yue, Jin Xu, Pengcheng Yin, Jinjing Luo and Hexin Wang et al.
Broadband-Printed Traveling-Wave Tube Based on a Staggered Rings Microstrip Line Slow-Wave Structure
Reprinted from: *Electronics* **2022**, *11*, 384, doi:10.3390/electronics11030384 . . . . **39**

Limin Sun, Hua Huang, Shifeng Li, Zhengbang Liu, Hu He and Qifan Xiang et al.
Investigation on High-Efficiency Beam-Wave Interaction for Coaxial Multi-Beam Relativistic Klystron Amplifier
Reprinted from: *Electronics* **2022**, *11*, 281, doi:10.3390/electronics11020281 . . . . **49**

Changqing Zhang, Pan Pan, Xueliang Chen, Siming Su, Bowen Song and Ying Li et al.
Design and Experiments of the Sheet Electron Beam Transport with Periodic Cusped Magnetic Focusing for Terahertz Traveling-Wave Tubes
Reprinted from: *Electronics* **2021**, *10*, 3051, doi:10.3390/electronics10243051 . . . . **61**

Kexin Ma, Jun Cai and Jinjun Feng
Investigation of a Miniaturized E-Band Cosine-Vane Folded Waveguide Traveling-Wave Tube for Wireless Communication
Reprinted from: *Electronics* **2021**, *10*, 3054, doi:10.3390/electronics10243054 . . . . **71**

Jinjing Luo, Jin Xu, Pengcheng Yin, Ruichao Yang, Lingna Yue and Zhanliang Wang et al.
A 340 GHz High-Power Multi-Beam Overmoded Flat-Roofed Sine Waveguide Traveling Wave Tube
Reprinted from: *Electronics* **2021**, *10*, 3018, doi:10.3390/electronics10233018 . . . . **81**

Guo Guo, Zhenlin Yan, Zhenzhen Sun, Jianwei Liu, Ruichao Yang and Yubin Gong et al.
Broadband and Integratable 2 × 2 TWT Amplifier Unit for Millimeter Wave Phased Array Radar
Reprinted from: *Electronics* **2021**, *10*, 2808, doi:10.3390/electronics10222808 . . . . **91**

Zheng Wen, Jirun Luo and Wenqi Li
Green's Functions of Multi-Layered Plane Media with Arbitrary Boundary Conditions and Its Application on the Analysis of the Meander Line Slow-Wave Structure
Reprinted from: *Electronics* **2021**, *10*, 2716, doi:10.3390/electronics10212716 . . . . **99**

Yijun Zhu, Yang Xie, Ningfeng Bai and Xiaohan Sun
Inverse Design of a Microstrip Meander Line Slow Wave Structure with XGBoost and Neural Network
Reprinted from: *Electronics* **2021**, *10*, 2430, doi:10.3390/electronics10192430 . . . . **111**

Hexin Wang, Shaomeng Wang, Zhanliang Wang, Xinyi Li, Tenglong He and Duo Xu et al.
Study of an Attenuator Supporting Meander-Line Slow Wave Structure for Ka-Band TWT
Reprinted from: *Electronics* **2021**, *10*, 2372, doi:10.3390/electronics10192372 . . . . **121**

Pengcheng Yin, Jin Xu, Lingna Yue, Ruichao Yang, Hairong Yin and Guoqing Zhao et al.
A New Method to Focus SEBs Using the Periodic Magnetic Field and the Electrostatic Field
Reprinted from: *Electronics* **2021**, *10*, 2118, doi:10.3390/electronics10172118 . . . . **131**

Shasha Li, Feng Zhang, Cunjun Ruan, Yiyang Su and Pengpeng Wang
A G-Band High Output Power and Wide Bandwidth Sheet Beam Extended Interaction Klystron Design Operating at TM_{31} with 2π Mode
Reprinted from: *Electronics* **2021**, *10*, 1948, doi:10.3390/electronics10161948 . . . . **139**

Xingchen Yang, Chaohai Du, Ziwen Zhang, Juanfeng Zhu, Tiejun Huang and Pukun Liu
Linearly Polarized High-Purity Gaussian Beam Shaping and Coupling for 330 GHz/500 MHz DNP-NMR Application
Reprinted from: *Electronics* **2021**, *10*, 1508, doi:10.3390/electronics10131508 . . . . **151**

About the Editors

Jinjun Feng

Jinjun Feng received the Bachelor Degree from Tsinghua University, Beijing, and the Master and Ph.D Degree from Beijing Vacuum Electronics Research Institute (BVERI). He has been with BVERI since 1990 and engaged in the research and development of millimeter wave space TWTs, W-Band TWTs and higher frequency devices using micro-fabrication, gyrotrons, field emission devices, cesium beam tubes etc. He is currently the Vice General-Director of BVERI. Dr Feng was the TPC Chair of the IVEC2015 in Beijing. He was a TPC Co-Chair of the 8th U.K., Europe, China Millimeter Waves and Terahertz Technology Workshop (UCMMT) in Cardiff, UK in 2015 and a General Co-Chair of the 9th UCMMT, in Tsingdao, China in 2016. He is a Fellow of Chinese Institute of Electronics (CIE), a Fellow of Institute of Engineer and Technology (IET), UK. He has been a member of IEEE EDS Vacuum Electronics Technical Committee since 2010. He was Chair of IEEE EDS Beijing Chapter (2011-2012) and Chair of IEEE Beijing Section (2014-2015), and is currently the Chair of IEEE China Council since 2016. He received the IEEE EDS John R. Pierce Award for Excellence in Vacuum Electronics in 2019.

Yubin Gong

Yubin Gong was born in Yantai, Shandong Province of China. He received his bachelor's degree in applied optics from Changchun University of Science and Technology in 1989, and his master's degree and philosophy doctor's degree in physical electronics from University of Electronic Science and Technology of China in 1991 and 1998, respectively. In 1991, he joined the Institute of High Energy Electronics, University of Electronic Science and Technology of China, and engaged in teaching and scientific research in the field of vacuum electronics. Now he is an outstanding professor in the School of Electronic Science and Engineering, University of Electronic Science and Technology of China, director of Chengdu Branch of National Key Laboratory of Microwave Vacuum Electronics, and leader of a vacuum electronics innovation research group. His research interests include high-power millimeter-wave and terahertz wave vacuum electron devices and their applications, terahertz devices combining vacuum and solid-state electronics, terahertz electromagnetic biological effects, and vortex electromagnetic wave biomedical imaging. He has published more than 400 academic papers, and has been invited to give more than 30 presentations at academic conferences. He has won the second-class prize of National Technical Invention and the first prize of Science and Technology Progress in Sichuan Province.

Chaohai Du

Chao-Hai Du (Senior Member, IEEE) received a B.S. degree in electronics from the University of Electronic Science and Technology of China, Chengdu, China, in 2005 and a Ph.D. degree in physical electronics from the University of Chinese Academy of Sciences, Beijing, China, in 2010. He is currently a Research Professor with Peking University, Beijing, China.

Adrian Cross

Adrian W. Cross was born in Hanover Germany in 1966. He received his B.Sc. degree (with honors) in physics and his Ph.D. degree from the University of Strathclyde, Glasgow, U.K., in 1989 and 1993, respectively. He joined the Atoms, Beams, and Plasmas Group, University of Strathclyde, in 1993 initially as a Research Fellow and then as a Lecturer in 2000, Senior Lecturer in 2003, Reader in 2006 and was promoted to Professor in 2014 with the Department of Physics, University of Strathclyde. From 2002 to 2007, he was an Engineering and Physical Science (EPSRC) UK Advanced Fellow and has been group leader since 2014. He has been involved in various aspects of research on gyrotrons, cyclotron autoresonance masers, free-electron lasers, superradiant sources, gyrotron travelling wave amplifiers, microwave pulse compression and plasma applications. More recently, he has primarily been concerned with research on THz radiation sources and pseudospark physics.

Preface to "High-Frequency Vacuum Electron Devices"

Vacuum electron devices at frequencies of millimeter waves and terahertz play highly important roles in the modern high-data rate and broadband communication systems, high-resolution detection and imaging, medical diagnostics, magnetically confined nuclear fusion, etc. For the fast motion velocity of electrons in the vacuum medium, they have the advantages of high power and high efficiency, as well as compactness, compared with other present radiation sources, such as solid-state devices.

We established the Special Issue of "High-Frequency Vacuum Electron Devices" with the aim of enhancing the exchange of research information on the theory, design, simulation, processes, and development of these devices to promote their applications, and to attract young researchers and engineers starting out in this important field, which is still vital on the basis of modern electronic science and information technology.

Jinjun Feng, Yubin Gong, Chaohai Du, and Adrian Cross
Editors

 MDPI

Editorial

High-Frequency Vacuum Electron Devices

Jinjun Feng [1,*], Yubin Gong [2,*], Chaohai Du [3,*] and Adrian Cross [4,*]

1 National Key Laboratory of Science and Technology on Vacuum Electronics, Beijing Vacuum Electronics Research Institute, Beijing 100015, China
2 School of Electronic Science and Engineering, University of Electronic Science and Technology of China, Chengdu 610054, China
3 School of Electronics Engineering and Computer Science, Peking University, Beijing 100871, China
4 Department of Physics, University of Strathclyde, Glasgow G1 1XQ, UK
* Correspondence: fengjinjun@tsinghua.org.cn (J.F.); ybgong@uestc.edu.cn (Y.G.); duchaohai@pku.edu.cn (C.D.); a.w.cross@strath.ac.uk (A.C.)

Citation: Feng, J.; Gong, Y.; Du, C.; Cross, A. High-Frequency Vacuum Electron Devices. *Electronics* **2022**, *11*, 817. https://doi.org/10.3390/electronics11050817

Received: 2 March 2022
Accepted: 3 March 2022
Published: 5 March 2022

Publisher's Note: MDPI stays neutral with regard to jurisdictional claims in published maps and institutional affiliations.

Vacuum electron devices at frequencies of millimeter waves and terahertz play highly important roles in the modern high-data rate and broadband communication system, high-resolution detection and imaging, medical diagnostics, magnetically confined nuclear fusion, etc. For the fast motion velocity of electrons in the vacuum medium, they have the advantages of high power and high efficiency, as well as compactness, compared with other present radiation sources, such as solid-state devices.

We established the Special Issue of "High-Frequency Vacuum Electron Devices" with the aim of enhancing the exchange of research information on the theory, design, simulation, processes, and development of these devices to promote their applications, and to attract young researchers and engineers starting out in this important field, which is still vital on the basis of modern electronic science and information technology.

There are many kinds of vacuum electronic RF power devices, including linear-beam devices, cross-field devices, and fast-wave devices. At high frequencies up to terahertz, klystrons, TWTs, BWOs, and gyrotrons are widely studied either for their high power, or for their broad instant or tuning bandwidth. In order to obtain high-quality performances at millimeter wave and terahertz frequencies, novel technologies and processes have emerged in the past decade, including microfabrication using MEMS and 3D printing, new diamond-related materials for windows and attenuators. At the same time, new slow-wave structures and resonant structures have also been studied, such as meta-structures, high-order mode operation and sheet electron beams, which are used to obtain high power; spurious depression; and the mitigation of manufacturing difficulties, specifically in high-frequency regimes. Revolutionary technologies in the components and parts of the devices, including cathodes, electron guns, I/O structures, magnetic focusing systems, and collectors, have played critical roles in the development of high-frequency vacuum electron devices.

This Special Issue consists of fifteen papers covering a broad range of topics related to the design, simulation, manufacturing, and testing of high-frequency vacuum devices with a wide range of frequencies up to 340 GHz, and devices including gyrotrons, TWTs, and EIKs, together with beam-forming and confining cathodes, slow-wave structures, and mode converters, etc.

High-frequency gyrotrons are core devices for Dynamic Nuclear Polarization Nuclear Magnetic Resonation (DNP-NMR) applications to significantly improve the sensitivity and resolution of high-field NMR in medical systems and scientific research. In [1], entitled "Linearly Polarized High-Purity Gaussian Beam Shaping and Coupling for 330 GHz/500 MHz DNP-NMR Application", from Beijing University, the design and calculation of a corrugated TE11-HE11 mode converter and a three-port directional coupler for a 330 GHz/500 MHz DNP-NMR system are proposed. The output mode of the mode converter presents a highly linear polarization of 98.8% at 330 GHz for subsequent low loss transmission. Gyrotron-TWTs have the advantages of high power and a broad bandwidth at a millimeter wave frequency band for long-distance accurate detection and high-resolution imaging. In [2] of this issue of "Design of a Dual-Mode Input Structure for K/Ka-Band Gyrotron TWT", the structure for the input coupling, which is composed of two different types of input couplers, with the coaxial input coupler for the Ka-band $TE_{2,1}$ mode and the Y-type input coupler for the K-band $TE_{1,1}$ mode, is described. The designed dual-mode input coupler has the advantages of a broad bandwidth and low loss and can be used effectively in dual-band Gyro-TWTs.

TWTs are widely investigated at high frequencies for their advantages of a high bandwidth, high power, and compactness for applications in high-date rate communication and high-resolution imaging. In this issue, there are eight papers dealing with the research and development of TWTs with novel integrated slow-wave structures (SWSs) and meander-line structures, as well as whole tubes. Three papers focus on meander-line SWSs but with varied supporting structures and materials. One paper [3] is based on the design of multiple Dielectric-Supported Ridge-Loaded Rhombus-Shaped Wideband Meander-Line Slow-Wave Structures for a V-Band TWT; the simulations show that the structure has a larger bandwidth, higher gain, more stable structure, and better heat dissipation ability. For high-power TWTs with meander-line SWSs, a staggered-ring micro-strip line (SRML) structure [4] is proposed and designed in the paper of "Broadband-Printed Traveling-Wave Tube Based on a Staggered Rings Micro-strip Line Slow-Wave Structure", and the input and output structures with micro-strip probes and transition sections are also shown. The particle-in-cell (PIC) simulation results indicate that the SRML TWT has a maximum output of 322 W at 32.5 GHz under a beam voltage of 9.7 kV and a beam current of 380 mA, and the output power is over 100 W in the frequency range of 27 GHz to 38 GHz.

Two new methods for the analysis of meander-line SWSs are presented in the issue. As we know, machine learning (ML) and deep learning (DL) are widely investigated and applied in many fields, while in one paper [5] in this Special Issue, ML and DL are introduced in the design of vacuum devices, where exact numerical simulation data are used in the training as a form of supervised learning to obtain the geometric dimensions. They are also used for the design of D-band meander-line TWTs with 160 GHz central frequency. Another paper [6] proposes a new method for solving the dyadic Green's functions (DGF) and scalar Green's functions (SGF) of multi-layered plane media, in which the DGF and SGF are considered to be suitable for arbitrary boundary conditions and for the electromagnetic analysis of complicated structures, including meander-line SWSs.

For other novel SWSs for millimeter wave structures, a cosine-vane folded waveguide [7] is used for a miniaturized E-Band TWT for wireless communication, and an over-moded flat-roofed sine waveguide is designed for high-power multi-beam TWTs. The E-band TWT has the properties of 9 W with a low voltage of 9 kV and compactness; this investigation is performed by the team from BVERI. For the higher frequency of 340 GHz, a flat-roofed sine waveguide [8] with over-mode operation and multiple electron beams is designed, and the PIC simulation results show that 50 w power can be obtained with a −3 dB bandwidth of 13 GHz.

Sheet electron beams used in vacuum devices can significantly increase the output power; there are two papers in this issue focusing on beam forming and transport, which is critical for the success of the whole tubes. In one paper [9], a periodic cusped magnet (PCM) is used for the design and evaluation of a G-Band TWT. A high beam-current density of 285 A/cm^2 and a voltage of 24.5 kV are successfully verified through a beamstick tube with over 81% transmission and a distance of 37.5 mm, which has compactness and light-weight beneficial properties. In another paper [10], a new electron beam method is proposed using both a periodic magnetic field and an electrostatic field (PM-E), which has the ability to resist the influence of the assembly error in the practical tubes, and the electric field can be conveniently changed to correct the deflection of the beam trajectory.

The integration of tubes for phase array systems has the advantages of high power and high efficiency. One paper [11] in this issue proposes the design of a Ka band using double parallel-connecting micro-strip meander-line SWSs, realizing power output through 2×2 ports. For each output port of one channel, the simulation results reveal that the output power can reach a high power of 566 W and a broad -3 dB bandwidth of 7 GHz.

Klystrons often have a higher power, higher gain and higher efficiency than TWTs, which is addressed two papers in this issue. One paper [12] performs an investigation of the high efficiency of a coaxial multi-beam relativistic klystron. PIC simulation is used to analyze the beam–wave interaction physical process and optimize the high-frequency parameters. The calculation shows that the klystron with 14 electron beams and a 4.2 kA bam current can deliver 1.02 GW power and 48.7% efficiency at 500 kV beam voltage. Another paper [13] is based on the G-Band klystron, which operates in high-order mode and in klystron TM31 mode, and a barbell six-gap cavity is selected. The interaction calculation shows that there is no risk of mode competition resulting from the big mode separation, and the power of 650 W with a 3 dB-bandwidth of 700 MHz can be obtained at 16.5 kV voltage and 0.5 A beam current.

Processing is critical for practical devices, ref. [14] of this special issue supplies a new structure for the assembly of the meander line SWS, in which an attenuator is employed to support the meander line on the bottom of the enclosure rather than welding them together on the sides. The three-dimension Particle-in-cell (PIC) simulation results show that with a 4.4-kV, 200-mA sheet electron beam, a maximum output power of 126 W is obtained at 38 GHz with electronic efficiency of 14.3%.

Finally, one of the most promising fields in vacuum electronics is the combination of microelectronics and modern 2D materials for high-performance electron emission sources and transistors. One paper [15] is based on vacuum transistors using carbon nanotubes (CNT) as electron sources. In the vacuum triode, multi-walled CNTs are used, with the principle of field-assisted thermal emission, which is fabricated using microfabrication technologies and has the advantages of improving the stability and uniformity of the devices. The experiment shows that the CNT transistor exhibits an ON/OFF current ratio as high as 104, and the surface of the emitters shows much lower gas molecule absorption than cold field emitters.

Funding: This research received no external funding.

Conflicts of Interest: The authors declare no conflict of interest.

References

1. Yang, X.; Du, C.; Zhang, Z.; Zhu, J.; Huang, T.; Liu, P. Linearly Polarized High-Purity Gaussian Beam Shaping and Coupling for 330 GHz/500 MHz DNP-NMR Application. *Electronics* **2021**, *10*, 1508. [CrossRef]
2. Ma, M.; Zhao, Q.; Mo, K.; Zheng, S.; Peng, L.; Lv, Y.; Feng, J. Design of a Dual-Mode Input Structure for K/Ka-Band Gyrotron TWT. *Electronics* **2022**, *11*, 432. [CrossRef]
3. Wang, Y.; Dong, Y.; Zhu, X.; Guo, J.; Xu, D.; Wang, S.; Gong, Y. Multiple Dielectric-Supported Ridge-Loaded Rhombus-Shaped Wideband Meander-Line Slow-Wave Structure for a V-Band TWT. *Electronics* **2022**, *11*, 405. [CrossRef]
4. Yang, R.; Yue, L.; Xu, J.; Yin, P.; Luo, J.; Wang, H.; Jia, D.; Zhang, J.; Yin, H.; Cai, J.; et al. Broadband-Printed Traveling-Wave Tube Based on a Staggered Rings Microstrip Line Slow-Wave Structure. *Electronics* **2022**, *11*, 384. [CrossRef]
5. Zhu, Y.; Xie, Y.; Bai, N.; Sun, X. Inverse Design of a Microstrip Meander Line Slow Wave Structure with XGBoost and Neural Network. *Electronics* **2021**, *10*, 2430. [CrossRef]
6. Wen, Z.; Luo, J.; Li, W. Green's Functions of Multi-Layered Plane Media with Arbitrary Boundary Conditions and Its Application on the Analysis of the Meander Line Slow-Wave Structure. *Electronics* **2021**, *10*, 2716. [CrossRef]
7. Ma, K.; Cai, J.; Feng, J. Investigation of a Miniaturized E-Band Cosine-Vane Folded Waveguide Traveling-Wave Tube for Wireless Communication. *Electronics* **2021**, *10*, 3054. [CrossRef]
8. Luo, J.; Xu, J.; Yin, P.; Yang, R.; Yue, L.; Wang, Z.; Xu, L.; Feng, J.; Liu, W.; Wei, Y. A 340 GHz High-Power Multi-Beam Overmoded Flat-Roofed Sine Waveguide Traveling Wave Tube. *Electronics* **2021**, *10*, 3018. [CrossRef]
9. Zhang, C.; Pan, P.; Chen, X.; Su, S.; Song, B.; Li, Y.; Lü, S.; Cai, J.; Gong, Y.; Feng, J. Design and Experiments of the Sheet Electron Beam Transport with Periodic Cusped Magnetic Focusing for Terahertz Traveling-Wave Tubes. *Electronics* **2021**, *10*, 3051. [CrossRef]

10. Yin, P.; Xu, J.; Yue, L.; Yang, R.; Yin, H.; Zhao, G.; Guo, G.; Liu, J.; Wang, W.; Gong, Y.; et al. A New Method to Focus SEBs Using the Periodic Magnetic Field and the Electrostatic Field. *Electronics* **2021**, *10*, 2118. [CrossRef]
11. Guo, G.; Yan, Z.; Sun, Z.; Liu, J.; Yang, R.; Gong, Y.; Wei, Y. Broadband and Integratable 2 $\times$ 2 TWT Amplifier Unit for Millimeter Wave Phased Array Radar. *Electronics* **2021**, *10*, 2808. [CrossRef]
12. Sun, L.; Huang, H.; Li, S.; Liu, Z.; He, H.; Xiang, Q.; He, K.; Fang, X. Investigation on High-Efficiency Beam-Wave Interaction for Coaxial Multi-Beam Relativistic Klystron Amplifier. *Electronics* **2022**, *11*, 281. [CrossRef]
13. Li, S.; Zhang, F.; Ruan, C.; Su, Y.; Wang, P. A G-Band High Output Power and Wide Bandwidth Sheet Beam Extended Interaction Klystron Design Operating at TM31 with 2π Mode. *Electronics* **2021**, *10*, 1948. [CrossRef]
14. Wang, H.; Wang, S.; Wang, Z.; Li, X.; He, T.; Xu, D.; Duan, Z.; Lu, Z.; Gong, H.; Gong, Y. Study of an Attenuator Supporting Meander-Line Slow Wave Structure for Ka-Band TWT. *Electronics* **2021**, *10*, 2372. [CrossRef]
15. He, Y.; Li, Z.; Mao, S.; Zhan, F.; Wei, X. A Vacuum Transistor Based on Field-Assisted Thermionic Emission from a Multiwalled Carbon Nanotube. *Electronics* **2022**, *11*, 399. [CrossRef]

MDPI

Article

Design of a Dual-Mode Input Structure for K/Ka-Band Gyrotron TWT

Mengshi Ma [1], Qixiang Zhao [1,*], Kunshan Mo [1], Shuquan Zheng [1], Lin Peng [1], You Lv [1] and Jinjun Feng [2]

1 School of Information and Communication, Guilin University of Electronic Technology, Guilin 541004, China; m1654259600@163.com (M.M.); mokunshan@126.com (K.M.); sqzheng1120@163.com (S.Z.); penglin@guet.edu.cn (L.P.); lvyoumail@yeah.net (Y.L.)
2 National Key Laboratory of Science and Technology on Vacuum Electronics, Beijing Vacuum Electronics Research Institute, Beijing 100015, China; fengjinjun@tsinghua.org.cn
* Correspondence: zxqi@guet.edu.cn; Tel.: +86-13-708-182-612

Abstract: A dual-band gyrotron traveling wave amplifier (Gyro-TWT) can reduce the size, cost, and weight of a transmitter in dual-band radar and communication systems. In this paper, a dual-mode input coupler for K/Ka dual-band gyrotron traveling wave amplifier (Gyro-TWT) is designed. This structure is composed of two different types of input couplers, one is the coaxial input coupler for the Ka-band $TE_{2,1}$ Gyro-TWT and the other is a Y-type input coupler for the K-band $TE_{1,1}$ Gyro-TWT. For reducing the backward wave of the $TE_{2,1}$ mode reflecting into the Y-type input coupler to influence the operating bandwidth, a Bragg reflector with a strong mode selective characteristic is inserted between these two couplers, which can make the reflection coefficient of the $TE_{2,1}$ mode better than −1 dB and the phase matched in the whole bandwidth, and the transmission coefficient of the $TE_{1,1}$ mode can reach better than −1 dB. Based on the simulation results, the −1 dB bandwidth of the Ka-band $TE^{\square}_{1,0}$-$TE^{\bigcirc}_{2,1}$ mode input coupler reaches 3.32 GHz and the −1 dB bandwidth of K-band $TE^{\square}_{1,0}$-$TE^{\bigcirc}_{1,1}$ mode input coupler reaches 3.15 GHz. The designed dual-mode input coupler has the advantages of broad bandwidth and low loss and can be well used in dual-band Gyro-TWTs.

Keywords: dual-band; Gyro-TWT; input coupler; coaxial cavity; Bragg reflector

Citation: Ma, M.; Zhao, Q.; Mo, K.; Zheng, S.; Peng, L.; Lv, Y.; Feng, J. Design of a Dual-Mode Input Structure for K/Ka-Band Gyrotron TWT. *Electronics* **2022**, *11*, 432. https://doi.org/10.3390/electronics11030432

Academic Editor: Gianluca Traversi

Received: 25 December 2021
Accepted: 27 January 2022
Published: 30 January 2022

1. Introduction

Gyrotron traveling wave amplifiers (Gyro-TWTs) utilize the relativistic electron cyclotron maser instability to realize a high-power microwave source with high conversion efficiency [1]. Compared with solid-state devices, such as diode and semiconductor devices, which are the main microwave source in communications, low-power radar, and the field of satellite communication [2,3], Gyro-TWTs are widely used in high-power and high-resolution radar, plasma heating and electronic countermeasure [4–7]. Because of using a nonresonant fast wave structure, the Gyro-TWT features the capabilities of high-power and broad bandwidth from millimeter to submillimeter, even to terahertz band. In recent years, significant progress has been made in the development of Gyro-TWTs that are even higher power and more stable. The National Tsing Hua University reported a Ka-band TE_{11} Gyro-TWT with off-axis electrons, which can produce 93 kW saturated peak output power at 70 dB stable gain and 26.5% efficiency with a 3 dB bandwidth of 8.6% [8]. Distributed wall losses are employed in the experiment to suppress the spurious oscillations. A Q-band TE_{01} Gyro-TWT with an output power of 152 kW at 41 dB saturated gain was developed in the University of Electronics Science and Technology of China [9]. High harmonic large orbit Gyro-TWT amplifiers can operate at higher beam currents and have good mode selection characteristics, therefore, producing higher power [10–12]. The Institute of Applied Physics (IAP) reported a second harmonic large orbit Gyro-TWT in the Ka-band. A spiral corrugated waveguide interaction circuit was adopted to suppress mode competition. The continuous wave power can reach 7.7 kW, the 3 dB bandwidth is 2.6 GHz, the gain is 26 dB, and the peak power reaches 180 kW [10]. Moreover, the University of Strathclyde reported a large orbit Gyro-TWT in W-band, the corresponding peak power is 3.4 kW, the gain and bandwidth are 37 dB and 5.8 GHz [12], respectively.

With the development of radar technology, advanced radar and communication systems are aiming to work in a dual-frequency band, which requires the amplifier providing the dual-band amplification to reduce in size, cost and weight in the radar transmitter system. Therefore, studying a Gyro-TWT operating at dual-band is very meaningful. As an important component of Gyro-TWTs, the input coupler that couples the EM radiation into the Gyro-TWT interaction cavity should be studied [13–19]. There are many kinds of input couplers, such as single-slit side-wall coupler, coaxial coupler, and couplers based on power splitter type. Most of the input couplers only can operate at single-mode and single frequency bands. The University of Electronic Science and Technology of China designed a Ku/Ka-band dual-frequency Gyro-TWT input coupler, but its −1 dB bandwidth is only 2 GHz [19].

In this paper, a dual-mode large bandwidth input coupler for K/Ka Gyro-TWT is theoretically designed. This structure is composed of two different types of input couplers, one is the coaxial input coupler for Ka-band $TE_{2,1}$ Gyro-TWT and the other is a Y-type input coupler for K-band $TE_{1,1}$ Gyro-TWT. For reducing the backward wave of the $TE_{2,1}$ mode reflecting into the Y-type input coupler to influence the operating bandwidth, a Bragg reflector with a strong mode selective characteristic is inserted between these two couplers. The −1 dB bandwidth of the Ka-band $TE^{\square}_{1,0}$-$TE^{\bigcirc}_{2,1}$ mode input coupler is broadened to 3.32 GHz with the center operating frequency of 19 GHz and the −1 dB bandwidth of the K-band $TE^{\square}_{1,0}$-$TE^{\bigcirc}_{1,1}$ mode input coupler can reach 3.15 GHz with the center operating frequency of 34 GHz. The designed input coupler could be well used in the dual-band Gyro-TWTs.

The organization of the paper is such that the single designs of coaxial and Y-type input couplers are discussed in Section 2. The Bragg reflector is also described in detail. The design of a dual-band input coupler is presented in Section 3, including the isolation between these two couplers. The summary is given in Section 4.

2. Design of Input Coupler

Two types of input couplers are discussed in this section, one is based on coaxial coupling [20], which converts the $TE^{\square}_{1,0}$ mode of the rectangular waveguide to the $TE^{\bigcirc}_{2,1}$ mode of circular waveguide in the Ka-band and the other is based on Y-shape input couplers [21], which can transform the $TE^{\square}_{1,0}$ to $TE^{\bigcirc}_{1,1}$ in the K-band.

2.1. Input Coupler Based on Coaxial Coupling Cavity

Figure 1 shows the schematic diagram of the coaxial input couple, where r_a is the inner radius of the coaxial resonator, r_b is the outer radius of the coaxial resonator, r_c is the radius of the cut-off waveguide. Generally, r_c is about 0.7–0.8 times the size of r, which is the output waveguide radius. Port 1 is the standard rectangular waveguide port where $TE^{\square}_{1,0}$ is feed into, port 4 is the output port where $TE^{\bigcirc}_{2,1}$ is transmitted to the interaction cavity, port 3 is the cut-off port to prevent the EM radiation into the gun section. When $TE^{\square}_{1,0}$ is coupled to the coaxial resonator cavity, the $TE_{4,1,1}$ mode is excited in the coupled cavity [22]. There are four coupling rectangular slits between the coaxial resonator and the cylindrical cavity. Therefore, the $TE_{4,1,1}$ mode can be coupled to $TE^{\bigcirc}_{2,1}$.

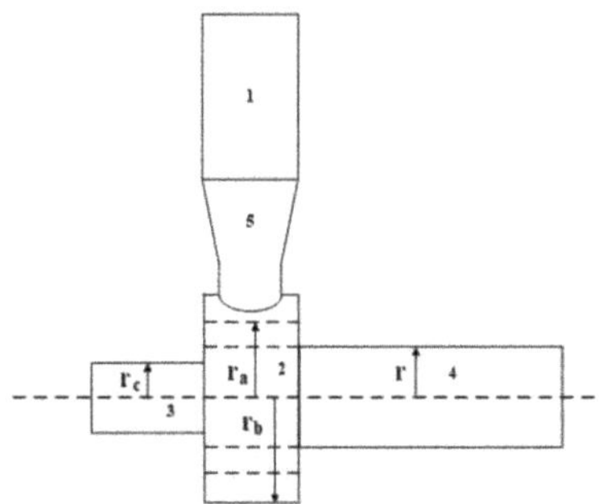

Figure 1. Schematic diagram of the coaxial input coupler.

It is necessary to study the Eigenvalue equation and propagation characteristics of the TE mode before designing the coupler, where the TE mode is only considered in the coaxial resonator. The longitudinal field component of the TE mode propagating in the axial waveguide is [13]:

$$H_z = K_c^2[C_m J_m(K_c r) + D_m Y_m(K_c r)] \sin m\varphi \, e^{-j\beta z} \tag{1}$$

where K_c is the cut-off wave number, m is the angular index of $TE_{m,n}$, $J_m(x)$ and $Y_m(x)$ are the mth order Bessel function and Neumann function, C_m and D_m are the mode amplitude.

From the boundary condition $\frac{\partial H_z}{\partial r}|_{\,r=a,b} = 0$, a homogeneous linear system of equations about C_m and D_m is obtained as:

$$\begin{cases} C_1 J'_m(K_c r_a) + D_1 Y'_m(K_c r_a) = 0 \\ C_2 J'_m(K_c r_b) + D_2 Y'_m(K_c r_b) = 0 \end{cases} \tag{2}$$

If there exists a non-zero solution of (2), the determinant is equal to zero, so the Eigenvalue equation of the $TE_{m,n}$ mode can be obtained as below

$$J'_m(\chi_{m,n}) Y'_m(\alpha\chi_{m,n}) - J'_m(\alpha\chi_{m,n}) Y'_m(\chi_{m,n}) = 0 \tag{3}$$

where $\chi_{m,n} = K_c r_a$ is the root of the Eigenvalue equation, $\alpha = r_b/r_a$ is the ratio of the inner and outer radius of the coaxial cavity. The propagation constant β_{mn} and resonance conditions of the TE_{mn} modes in coaxial waveguides are obtained as shown in (4) and (5):

$$\beta_{mn} = \sqrt{K^2 - K_c^2} = \sqrt{\omega^2 \varepsilon\mu - (\chi_{mn}/r_a)^2} \tag{4}$$

$$\beta_{mn} L = p\pi (p = 1, 2, 3 \ldots) \tag{5}$$

where L is the length of the coaxial resonator. By solving the Eigenvalue Equation (3) numerically, the relationship between the eigenvalue and the ratio of the inner and outer radius of the coaxial waveguide is obtained.

As shown in Figure 2, the Eigenvalue reduces with the rise of the ratio. To increase the coupling between the inner and outer cavity and reduce the size of the coupler, the thickness of the coupling silt is always 0.3~0.5 mm. Based on the results from a beam–wave interaction and the results in Figure 2, the geometric configuration of the input coupler is preliminary determined. Then, with the help of CST (in the simulation, the resolution is set as −40 dB), the structure of the input coupler is optimized and the field distribution at the frequency of 34 GHz is plotted in Figure 3. It is obvious that $TE_{2,1}^{\bigcirc}$ is excited at the circular cavity.

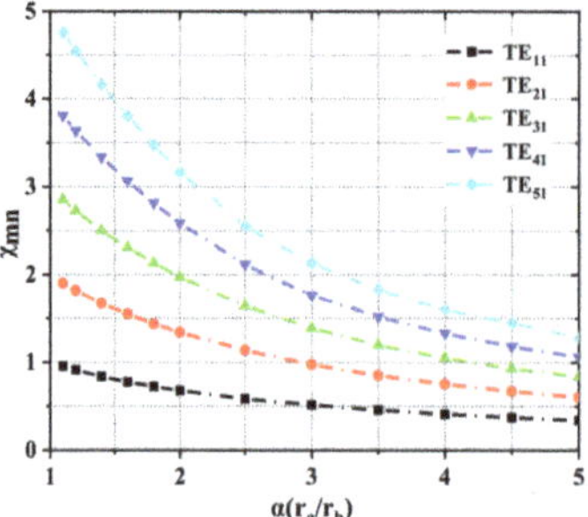

Figure 2. The relationship between the Eigenvalue root $\chi_{m,n}$ and the inner coaxial cavity and outer coaxial cavity radii ratio a.

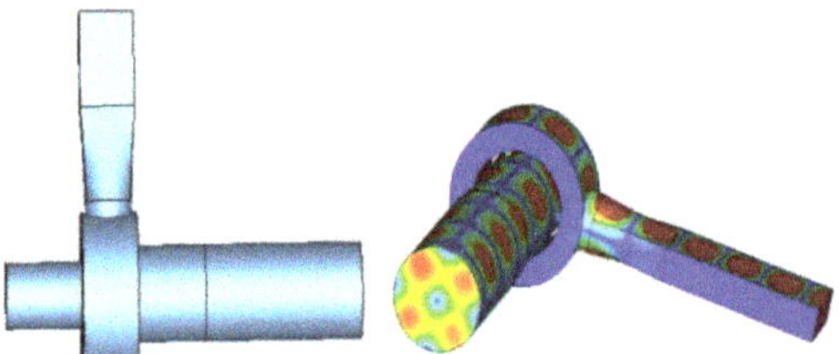

Figure 3. Coupler structure and field distribution diagram.

The influence of the width of the coupling rectangular gap on the transmission and reflection of the input coupler is plotted in Figure 4. It is found that the reflection coefficient increases slightly with the rise of the coupling gap width. Whereas the transmission is almost unchanged in the frequency range of 33–38 GHz. Figure 5 shows the influence of the coupling rectangular gap length on the performance of the input coupler. It can be seen that the transmission remains unchanged in the frequency range of 33–38 GHz when the gap length is increased. From the above analysis, the −1 dB bandwidth of the transmission can reach 4.30 GHz, and the reflection in the bandwidth are all below −15 dB. For reducing the reflection between the input rectangular and the coaxial cavity, a smooth variation section with a length of 10 mm is designed.

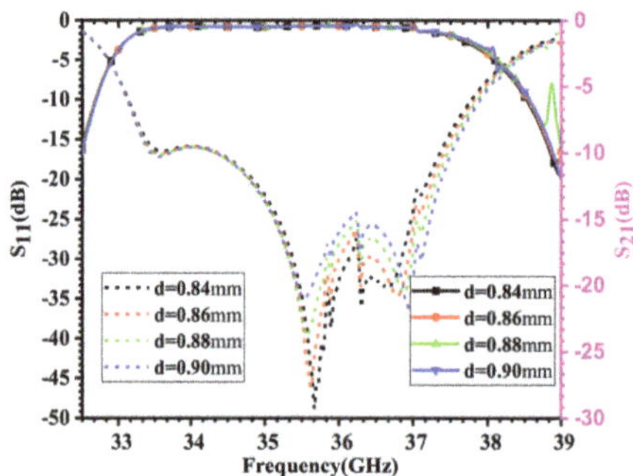

Figure 4. The influence of the coupling rectangle gap width on transmission and reflection of the input coupler.

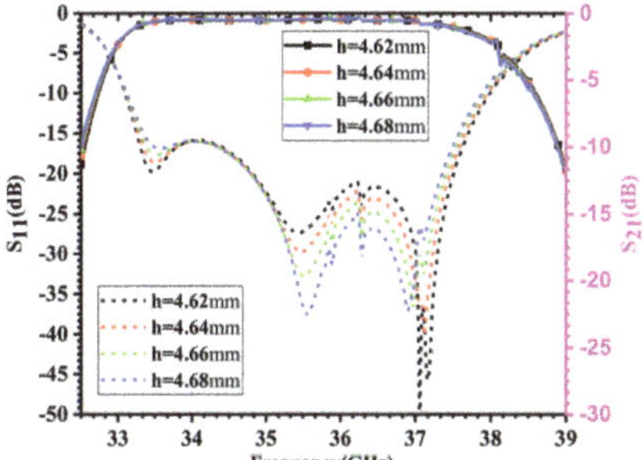

Figure 5. The influence of the coupling rectangle gap length on transmission and reflection of the input coupler.

2.2. Input Coupler Based on Side Wall COUPLING Mode

Mode $TE^{\bigcirc}_{1,1}$ is selected as the operating mode for Gyro-TWT in the K-band. We adopt the input coupler based on sidewall coupling, as shown in Figure 6. As we know, the circularly polarized TE_{11} mode can be synthesized from two same amplitude linearly polarized TE_{11} modes with a 90-degree phase difference [21]. Therefore, the coupler has two rectangular branches to generate two linearly polarized TE_{11} modes. The simulated electric field is shown in Figure 6. It is shown that $TE^{\bigcirc}_{1,1}$ can be excited in the circular cavity. The input rectangular waveguide is the waveguide standard BJ180. The circular waveguide is 5.1 mm, which is the same as the radius of the output cavity of the coaxial coupler. In addition, a tapered geometry with a length of 4 mm is used between the output rectangular waveguide and the circular waveguide to effectively reduce the reflection. The transmission and reflection are plotted in Figure 7. The frequency range of −1 dB bandwidth can reach 17.5~21.4 GHz, the bandwidth is 3.9 GHz.

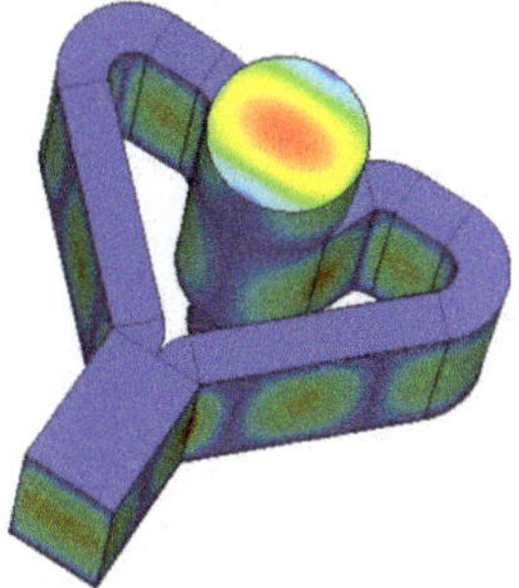

Figure 6. Geometric structure of the Y-type input coupler.

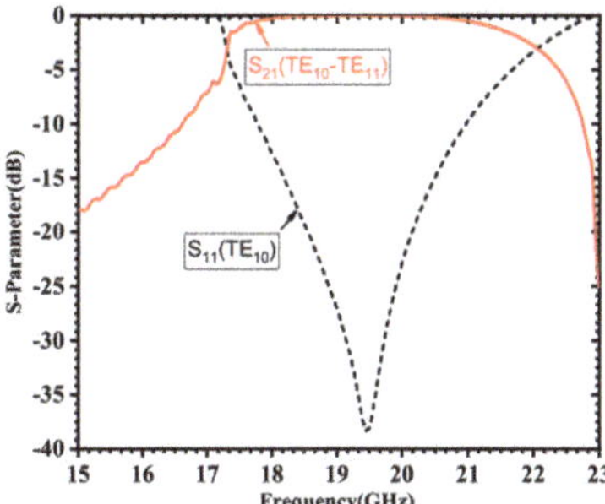

Figure 7. Simulation results of the Y-type input coupler.

2.3. Bragg Reflector

The cut-off waveguide is a general solution to improve the transmission coefficient for the input coupler. The only requirement is the radius of the cut-off waveguide needs to be sufficiently small. Thus, the radius of the cut-off waveguide in the coaxial cavity is 3.50 mm, which is smaller than the radius of the output cavity. To solve this problem, a Bragg reflector with a strong mode selection feature is adopted to connect the output end of the Y-type input coupler and the cut-off port of the coaxial input coupler [23–26]. The function of the Bragg reflector is to allow the $TE_{1,1}^{\bigcirc}$ mode transmission with no reflection and prevent the $TE_{2,1}^{\bigcirc}$ mode transmitting into the Y-type coupler. The Bragg reflectors can be constructed in a variety of ways, among the various axial periodic structures, the simplest structure is the periodic rectangular-corrugation waveguide, which includes two circular waveguide sections with different radii in one period. According to the Bragg resonance conditions given in [23] ($2k = k_b$, where k is the axial propagation constant of the wave, $k_b = 2\pi/l$, l is the corrugation period), the geometric structure of the Bragg reflector is determined and a strong mode-selective reflection that scatters the incident wave coherently into a backward wave can be achieved. It was shown that the bandwidth of the Bragg reflector can be improved by varying the corrugation profile. As shown in Figure 8, a seven-section reflector was designed to operate in the frequency range of 33~38 GHz and 17.5~21.4 GHz. The minimum radius of the reflector R_{in} is the same as the radius of the interaction zone. Based on the joint simulation of CST and MATLAB, the length of each section is shown in Table 1. The outer radius R_{out} is 11.63 mm. Figure 9 shows the reflection coefficient and phase of the $TE_{2,1}^{\bigcirc}$ mode. The results show that the reflector achieves total reflection in the frequency range of 33~38 GHz, the phase spread can be even smaller, and the transmission S_{21} in the frequency range of 17.5~21.4 GHz is close to 0 dB and the corresponding reflection S_{11} is below −10 dB. The designed reflector satisfies the dual-band input coupler.

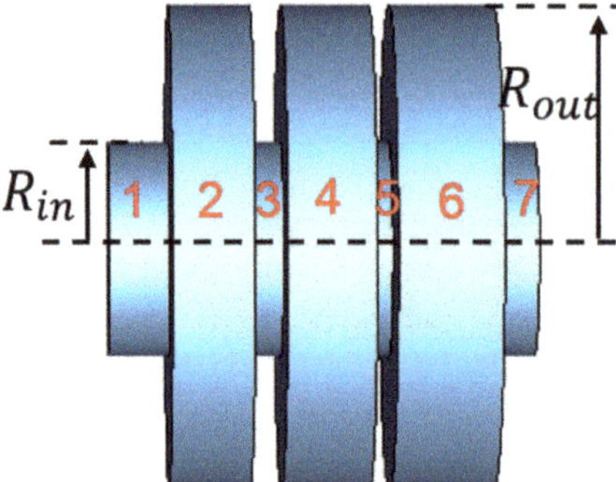

Figure 8. Structure diagram of the Bragg reflector.

Table 1. Dimensions of the Bragg reflector.

Section Number	Length (mm)
1	3
2	4
3	1.6
4	4.3
5	1.5
6	4.3
7	2

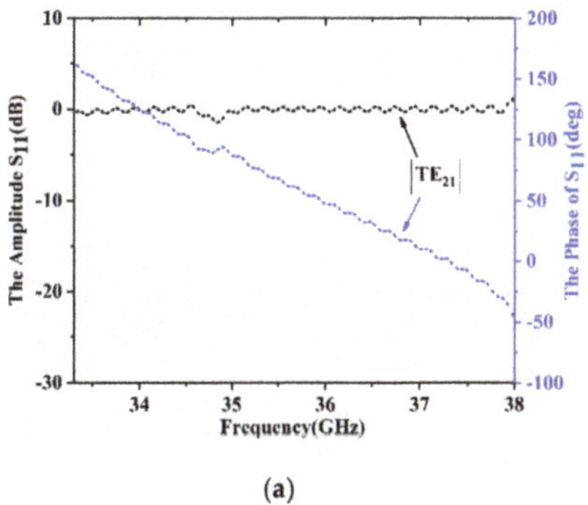

(a)

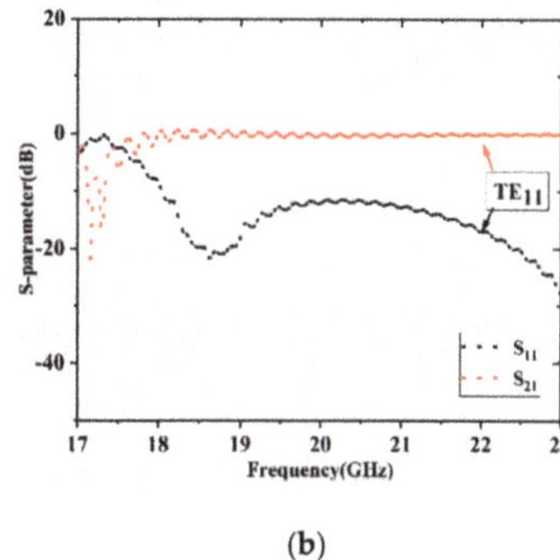

(b)

Figure 9. (**a**) Reflection amplitude (the black-dot curve) and phase (the blue-dot curve) of S_{11} of $TE^{\bigcirc}_{1,1}$ mode, (**b**) the reflection and transmission of $TE^{\bigcirc}_{1,1}$ mode.

3. Simulation of Dual-Mode Input Couplers

Based on the front studies, a dual-mode input coupler for K/Ka dual-band Gyro-TWT is designed, as shown in Figure 10. Port 1 is the input port for $TE^{\square}_{1,0}$ in the K-band, port 2 is the input port for $TE^{\square}_{1,0}$ in the Ka-band, port 3 is the output port with a radius of 5.1 mm, Section 4 is the Bragg reflector, and port 5 is the cut-off port. Figure 11 shows the electric field distribution of the input coupler when the operating mode is $TE^{\bigcirc}_{1,1}$ at a frequency of 18 GHz and the $TE^{\bigcirc}_{2,1}$ at a frequency of 34 GHz. It can be seen that the electric field is blocked in the Bragg reflector and no field is transmitting into the Y-type input coupler when the dual-mode input coupler is operating at $TE^{\bigcirc}_{2,1}$, and the Bragg reflector has no influence on the transmission of $TE^{\bigcirc}_{1,1}$. Figures 12 and 13 plot the S-parameter and phase of the designed dual-mode input coupler in the K/Ka-band. It is shown that the reflection coefficient in the frequency range of 18.3~21 GHz is below −10 dB when the output mode is $TE^{\bigcirc}_{1,1}$, the −1 dB bandwidth of transmission can reach 3.15 GHz. Meanwhile, the reflection coefficient in the frequency range of 33.7~37.2GHz is below −10 dB when the output mode is $TE^{\bigcirc}_{2,1}$, and the −1 dB bandwidth is 3.32 GHz. The comparison between the presented input coupler and that in [19] is shown in Table 2. It can be seen that the input coupler bandwidth is relatively broadened to wide. The designed input coupler can be used as a dual-band large bandwidth Gyro-TWT.

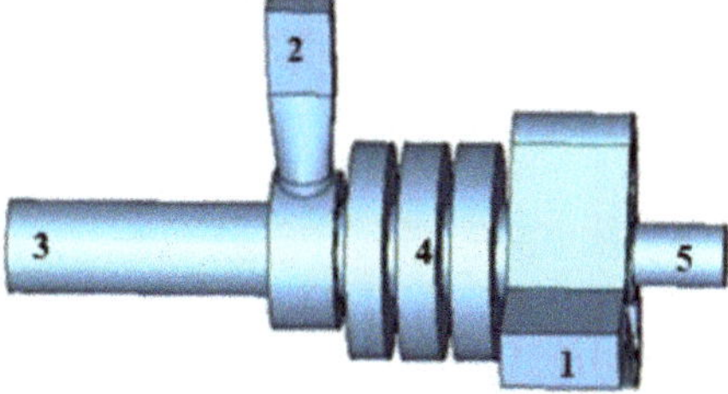

Figure 10. Structure diagram of dual-frequency input coupler (1 is the input port for $TE^{\square}_{1,1}$ in K-band, 2 is the input port for $TE^{\square}_{1,1}$ in Ka-band, 3 is the output port, 4 is the Bragg reflector, 5 is the cut-off port).

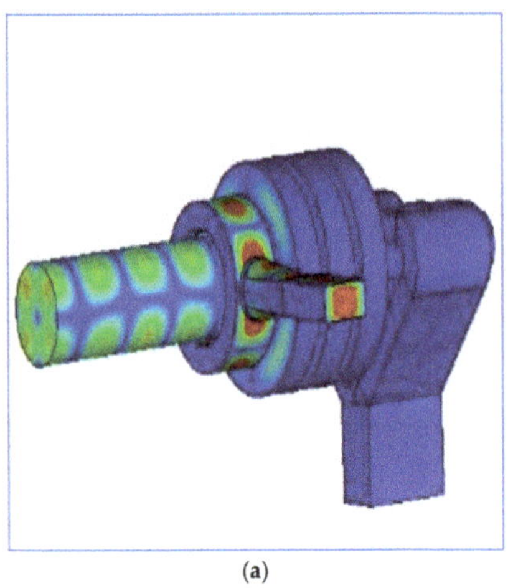

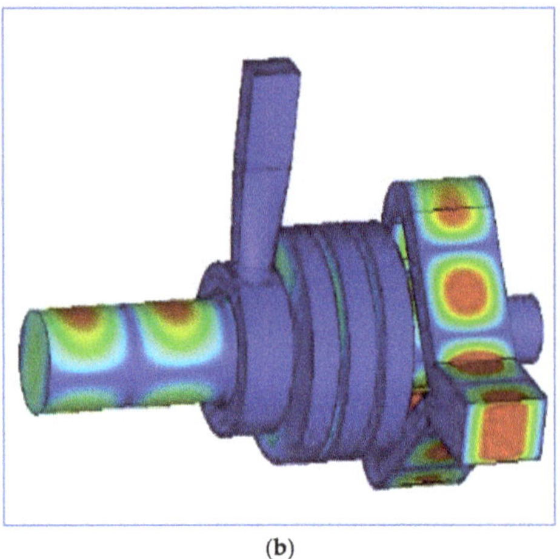

(a) (b)

Figure 11. The electric field diagram of input coupler, (**a**) the output mode is $TE_{1,1}^{\bigcirc}$, (**b**) the output mode is $TE_{1,1}^{\bigcirc}$.

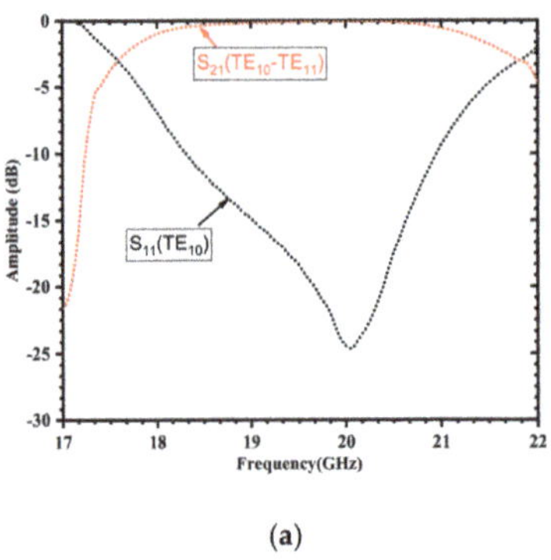

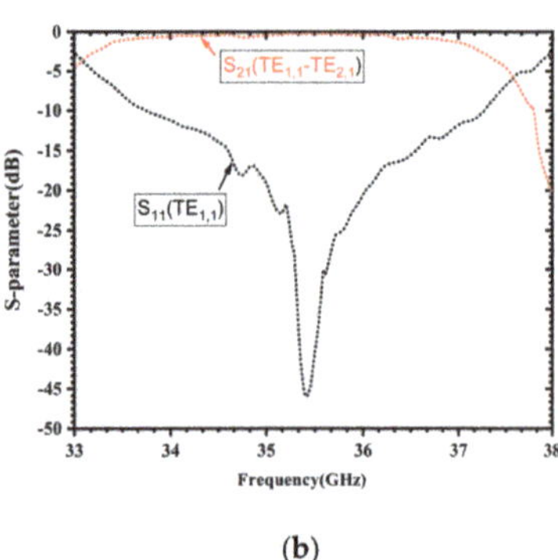

(a) (b)

Figure 12. The S-parameter for the designed dual-mode input coupler, (**a**) the S-parameter when the output mode is $TE_{1,1}^{\bigcirc}$, (**b**) the S-parameter when the output mode is $TE_{2,1}^{\bigcirc}$.

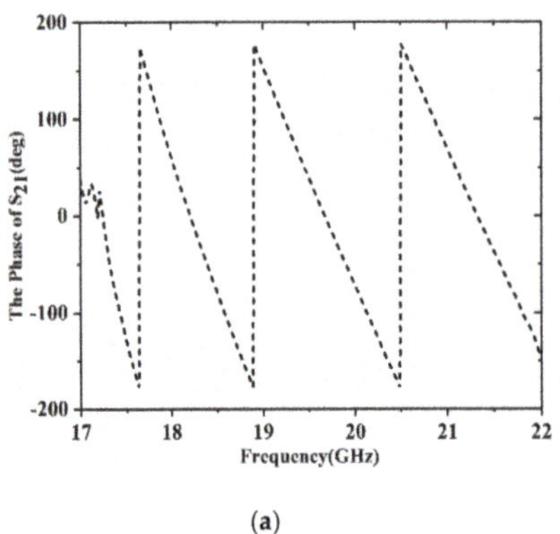

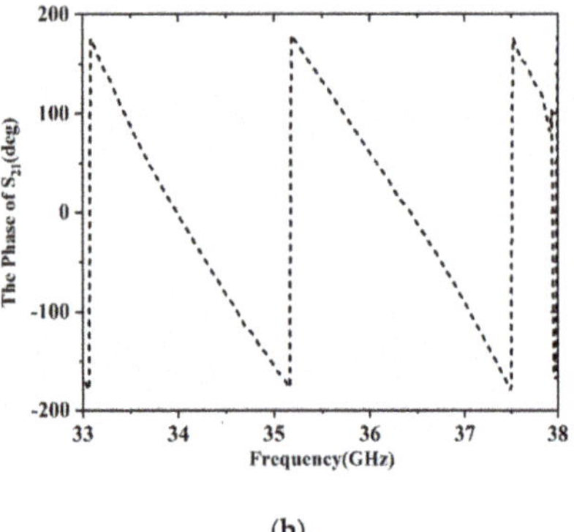

(a) (b)

Figure 13. The phase of S_{21} for the designed dual-mode input coupler, (**a**) the phase of S_{21} when the output mode is $TE_{1,1}^{\bigcirc}$, (**b**) the phase of S_{21} when the output mode is $TE_{2,1}^{\bigcirc}$.

Table 2. The dual-band input coupler parameters.

Parameters	This Article	Reference [19]
Operating Frequency	19/34GHz	17/33GHz
Operating Mode	$TE^{\square}_{1,0}$-$TE^{\bigcirc}_{2,1}$ @ Ka-band and $TE^{\square}_{1,0}$-$TE^{\bigcirc}_{1,1}$ @ K-band	$TE^{\square}_{1,0}$-$TE^{\bigcirc}_{0,1}$ @ Ka-band and $TE^{\square}_{1,0}$-$TE^{\bigcirc}_{1,1}$ @ K-band
−1 dB Bandwidth	3.15 GHz @ K-band,3.32 GHz@ Ka-band	1.9 GHz @ K-band, 2 GHz@ Ka-band

Figure 14 plots the isolation between the two input rectangular waveguides when the designed dual-mode input coupler is operating at K-band and Ka-band, respectively. When the coupler is operating in the K-band, the isolation of the $TE^{\bigcirc}_{1,1}$ mode is below −40 dB in the 17~22GHz frequency range. When operating in the Ka-band, the isolation of the $TE^{\bigcirc}_{2,1}$ mode is below −20 dB in the frequency range of 33~38 GHz. Through these analyses, it is proved that the designed input coupler has good properties for single-mode operating.

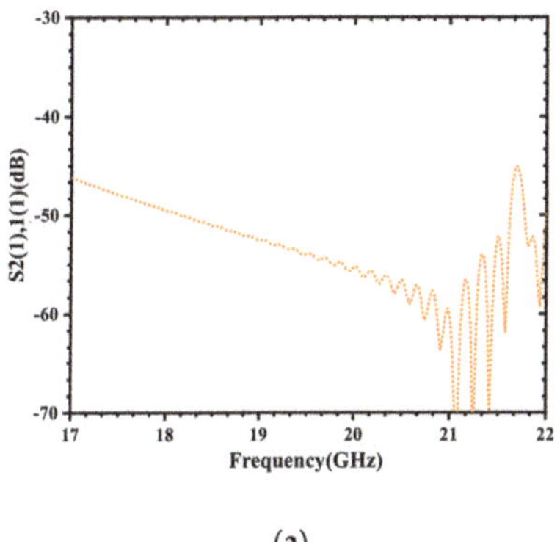

(**a**)

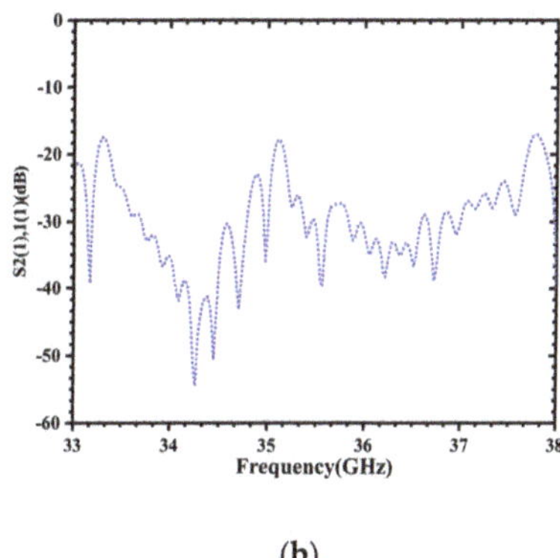

(**b**)

Figure 14. The isolation between the input rectangular (**a**) port 1 and (**b**) port 2 (ports 1 and 2 are labeled in Figure 10).

4. Conclusions

Dual-band Gyro-TWTs can be used in new radars, communication systems and other fields to achieve cross-band operation requirements. In this paper, a Gyro-TWT input coupler for dual-band operation is presented. The coupler realizes the transition from the rectangular waveguide mode to the circular waveguide mode in different wavebands. The designed structure is composed of two different types of input couplers, one is the coaxial input coupler for the Ka-band $TE_{2,1}$ Gyro-TWT and the other is a Y-type input coupler for the K-band $TE_{1,1}$ Gyro-TWT. For reducing the reflection wave from $TE_{2,1}$ reflecting into the Y-type input coupler to influence the operating bandwidth, a Bragg reflector is inserted to connect these two couplers. Through optimization, the −1 dB bandwidth of the dual-band input coupler can achieve 3.15 GHz in the K-band with the output mode of $TE^{\bigcirc}_{1,1}$ and 3.32 GHz in the Ka-band with the output mode of $TE^{\bigcirc}_{2,1}$. Meanwhile, the isolation degree of the two input rectangular ports is below −20 dB. The designed dual-mode input coupler can be well used for dual-band Gyro-TWTs.

Author Contributions: Conceptualization, M.M. and Q.Z.; methodology, M.M., L.P.; resources, S.Z., K.M., Y.L. and J.F.; data curation, Q.Z.; formal analysis, Q.Z., J.F.; writing—original draft preparation, M.M. and Q.Z.; writing—review and editing, Q.Z., J.F.; funding acquisition, Q.Z., J.F. All authors have read and agreed to the published version of the manuscript.

Funding: This research was funded by the National Natural Science Foundation of China: 62001131, the Dean Project of Guangxi Key Laboratory of Wireless Broadband Communication and Signal Processing Grant Nos: GXKL06190102, the Guangxi Natural Science Foundation Project: 2019GXNSFBA245066, and the Guangxi Science and Technology Base and Talent Special Project: AD19245042.

Conflicts of Interest: The authors declare no conflict of interest.

References

1. Chu, K.R.; Lin, A.T. Gain and bandwidth of the Gyro-TWT and CARM amplifiers. *IEEE Trans. Plasma Sci.* **1988**, *16*, 90–104. [CrossRef]
2. Li, H.H.; Wang, K.S.; Zhao, J.X.; Xiao-xin, L.I.A.N.G.; Yue-peng, Y.A.N. A design of Ka-band power amplifier based on 0. 15 μm GaAs pHEMT process. *Microelectron. Comput.* **2021**, *38*, 17–21.
3. Hosseinzadeh, N.; Medi, A. Wideband 5 W Ka-Band GaAs Power Amplifier. *IEEE Microw. Wirel. Compon. Lett.* **2016**, *26*, 1–3. [CrossRef]
4. Chu, K.R. Overview of Research on the gyrotron traveling-wave amplifier. *IEEE Trans Plasma Sci.* **2002**, *30*, 903–908.
5. Baik, C.-W.; Jeon, S.-G.; Kim, D.-H.; Sato, N.; Yokoo, K.; Park, G.-S. Third-harmonic frequency multiplication of a two-stage tapered gyrotron TWT amplifier. *IEEE Trans. Electron Devices* **2005**, *52*, 829–838. [CrossRef]
6. Linde, G.J.; Ngo, M.T.; Danly, B.G.; Cheung, W.J.; Gregers-Hansen, V. WARLOC: A high-power coherent 94 GHz radar. *IEEE Trans. Aerosp. Electron. Syst.* **2008**, *44*, 1102–1117. [CrossRef]
7. Paoloni, C.; Gamzina, D.; Letizia, R.; Zheng, Y.; Luhmann, N.C., Jr. Millimeter wave traveling wave tubes for the 21st Century. *J. Electromagn. Waves Appl.* **2021**, *35*, 567–603. [CrossRef]
8. Chu, K.R.; Chen, H.Y.; Hung, C.L.; Chang, T.-H.; Barnett, L.; Chen, S.-H.; Yang, T.-T.; Dialetis, D.J. Theory and experiment of ultrahigh-gain gyrotron traveling wave amplifier. *IEEE Trans. Plasma Sci.* **1999**, *27*, 391–404.
9. Yan, R.; Luo, Y.; Liu, G.; Pu, Y. Design and experiment of a Q-band gyro-TWT loaded with lossy dielectric. *IEEE Trans. Electron. Devices* **2012**, *59*, 3612–3617. [CrossRef]
10. Samsonov, S.V.; Gachev, I.G.; Denisov, G.G.; Bogdashov, A.; Mishakin, S.V.; Fiks, A.S.; Soluyanova, E.A.; Tai, E.M.; Dominyuk, Y.V.; Levitan, B.A.; et al. Ka-Band gyrotron traveling-wave tubes with the highest continuous-wave and average power. *IEEE Trans. Electron. Devices* **2014**, *61*, 4264–4267. [CrossRef]
11. Harriet, S.B.; McDermott, D.B.; Gallagher, D.A.; Luhmann, N. Cusp Gun TE21 Second-Harmonic Ka-Band Gyro-TWT Amplifier. *IEEE Trans. Plasma Sci.* **2002**, *30*, 909–914. [CrossRef]
12. He, W.; Donaldson, C.R.; Zhang, L.; Ronald, K.; Phelps, A.D.R.; Cross, A.W. Broadband amplification of low terahertz signals using axis-encircling electrons in a helically corrugated interaction region. *Phys. Rev. Letts.* **2017**, *119*, 184801. [CrossRef] [PubMed]
13. Xu, Y.; Xiong, C.D.; Yong, L.; Jianxun, W.; Ran, Y.; Youlei, P.; Wang, H.; Li, H. Design of broad-band input coupler of Ka-band TE_{01} mode gyro-TWT. *Chin. J. Vac. Sci. Technol.* **2012**, *32*, 208–213.
14. Xiong, W.J.; Wang, L.; Luo, Y.; Guo, L. Improved design of input and output structures of W band gyro-TWT. *High Power Laser Part. Beams* **2013**, *25*, 693–698. [CrossRef]
15. Wang, Q.S.; Huey, H.E.; McDermott, D.B.; Hirata, Y.; Luhmann, N. Design of a W-band Second-harmonic TE_{02} gyro-TWT amplifier. *IEEE Trans. Plasma Sci.* **2000**, *28*, 2232–2238. [CrossRef]
16. McDermott, D.B.; Song, H.H.; Hirata, Y.; Lin, A.T.; Barnett, L.R.; Chang, T.H.; Hsu, H.-L.; Marandos, P.S.; Lee, J.; Chu, K.R.; et al. Design of a W-band TE_{01} mode gyrotron traveling-wave amplifier with high power and broad-band capabilities. *IEEE Trans. Plasma Sci.* **2003**, *30*, 894–902. [CrossRef]
17. Liu, G. Input coupler design for Ka band gyrotron TWT. In Proceedings of the 2016 IEEE International Vacuum Electronics Conference (IVEC), Monterey, CA, USA, 19–21 April 2016; pp. 1–2.
18. Zhang, L.; He, W.; Donaldson, C.; Garner, J.R.; McElhinney, P.; Cross, A.W. Design and Measurement of a Broadband Sidewall Coupler for a W-Band Gyro-TWA. *IEEE Trans. Microw. Theory Tech.* **2015**, *63*, 3183–3190. [CrossRef]
19. Sun, M. *Research on the Transmission Link of Gyro-TWT Input and Output System [D]*; University of Electronic Science and Technology of China: Chengdu, China, 2019; pp. 34–65.
20. Xu, Y.; Li, Y.; Wang, J.X.; Jiang, W.; Liu, G.; Luo, Y.; Li, H. Design and experiment of a high power and broadband Ku-Band TE11 mode Gyro-TWT. *IEEE Trans. Electron. Devices* **2018**, *66*, 1559–1566. [CrossRef]
21. Yu, C.F.; Chang, T.H. High-performance circular TE01-mode converter. *IEEE Trans. Microw. Theory Tech.* **2005**, *53*, 3794–3798.
22. Collin, R.E. *Field Theory of Guided Waves*; Wiley Interscience: New York, NY, USA, 1990; pp. 415–420.
23. Chong, C.K.; McDermott, D.B.; Razeghi, M.M.; Luhmann, N.C.; Pretterebner, J.; Wagner, D.; Thumm, M.; Caplan, M.; Kulke, B. Bragg reflectors. *IEEE Trans. Plasma Sci.* **1992**, *20*, 393–402. [CrossRef]

24. Peskov, N.Y.; Ginzburg, N.S.; Kaminskii, A.A.; Kaminskii, A.K.; Sedykh, S.N.; Sergeev, A.P.; Sergeev, A.S. High-efficiency narrow-band free-electron maser using a Bragg cavity with a phase discontinuity in the ripples. *Tech. Phys. Lett.* **1999**, *25*, 429–432. [CrossRef]
25. Emile, D.R. Innovative corrugated transmission line for Terahertz wave-guiding. In Proceedings of the 2011 International Conference on Infrared, Millimeter, and Terahertz Waves, Houston, TX, USA, 2–7 October 2011; pp. 1–2.
26. Wenzel, H.; Guther, R.; Shams-Zadeh-Amiri, A.M.; Bienstman, P. A comparative study of higher order Bragg gratings: Coupled-mode theory versus mode expansion modeling. *IEEE J. Quantum Electron.* **2006**, *42*, 64–70. [CrossRef]

Article

A Vacuum Transistor Based on Field-Assisted Thermionic Emission from a Multiwalled Carbon Nanotube

Yidan He †, Zhiwei Li †, Shuyu Mao, Fangyuan Zhan and Xianlong Wei *

Key Laboratory for the Physics and Chemistry of Nanodevices, School of Electronics, Peking University, Beijing 100871, China; yidan@stu.pku.edu.cn (Y.H.); lzw111@pku.edu.cn (Z.L.); 1700012814@pku.edu.cn (S.M.); zhanfangy@163.com (F.Z.)

* Correspondence: weixl@pku.edu.cn

† These authors contributed equally to this work.

Abstract: Vacuum triodes have been scaled down to the microscale on a chip by microfabrication technologies to be vacuum transistors. Most of the reported devices are based on field electron emission, which suffer from the problems of unstable electron emission, poor uniformity, and high requirement for operating vacuum. Here, to overcome these problems, a vacuum transistor based on field-assisted thermionic emission from individual carbon nanotubes is proposed and fabricated using microfabrication technologies. The carbon nanotube vacuum transistor exhibits an ON/OFF current ratio as high as 10^4 and a subthreshold slope of ~4 $V \cdot dec^{-1}$. The gate controllability is found to be strongly dependent on the distance between the collector electrodes and electron emitter, and a device with the distance of 1.5 μm shows a better gate controllability than that with the distance of 0.5 μm. Benefiting from field-assisted thermionic emission mechanism, electric field required in our devices is about one order of magnitude smaller than that in the devices based on field electron emission, and the surface of the emitters shows much less gas molecule absorption than cold field emitters. These are expected to be helpful for improving the stability and uniformity of the devices.

Keywords: vacuum transistors; field-assisted thermionic emission; carbon nanotubes; gate controllability

Citation: He, Y.; Li, Z.; Mao, S.; Zhan, F.; Wei, X. A Vacuum Transistor Based on Field-Assisted Thermionic Emission from a Multiwalled Carbon Nanotube. *Electronics* **2022**, *11*, 399. https://doi.org/10.3390/electronics11030399

Academic Editor: Yahya M. Meziani

Received: 28 December 2021
Accepted: 25 January 2022
Published: 28 January 2022

1. Introduction

Vacuum tubes emerged in the early 20th century and were the central of the original electronic devices [1]. However, solid-state devices took over their roles in most areas in the past 60 years because of the advantages of integrability, miniaturization, lower power consumption, reduced costs, etc. Recently, vacuum transistors, miniature vacuum triodes fabricated on a chip by microfabrication technologies, have rekindled many researchers' interest because of the advantages associated with vacuum devices. Vacuum as a medium for electron transport is more immune to radiation damage than conventional semiconductors. Thus, vacuum devices are stable under harsh environment [2]. Furthermore, the velocity of electrons transporting in a vacuum is higher than that in semiconductors, because electrons in a vacuum are free of scattering with a theoretical velocity approaching 3×10^8 $m \cdot s^{-1}$. Meanwhile, vacuum devices are more reliable and efficient than solid-state devices for high-power and high-frequency devices [3]. Up to now, the functionality of vacuum devices has been demonstrated in a wide range of applications, such as deep space communications [2], premier sound system [4], high-frequency and high-power devices [5], terahertz laser [6] and military defense. Combining vacuum triodes and microfabrication technologies leads to the creation of a new area called "vacuum transistors", which is expected to possess the advantages of both conventional vacuum devices and solid-state devices, such as high carrier velocity, reliable performances in high temperature and extreme environment, miniaturization and easy integration [7].

In recent years, as many researchers have put their efforts into this field, various vacuum transistors fabricated by microfabrication technologies have been reported [7–17]. Miniaturization of vacuum triodes can lead to higher integration, lower working voltages and lower power consumption. For instance, Shruti Nirantar et al. proposed a semiconductor-free field emission nanoscale channel transistor, where the gap between field emission electrodes is about 35 nm [8]. As the gap is less than the mean free path of electrons in the air pressure (~60 nm), electrons encounter fewer collisions with air molecules even in air pressure [9]. Jin-Woo Han et al. fabricated a surround gate nanoscale vacuum channel transistor by using ion implantation and ion etching, achieving a low operating voltage (<5 V) and good immunity to various types of radiation [10]. In addition, Jin-Woo Han et al. reported vertical surround-gate nanoscale vacuum transistors that can be fabricated on silicon carbide wafers with stable electron emission and long-term stability of emitters [7]. The extended gate structure enhances the gate-to-emitter controllability and reduces the leakage current. A fully integrated, on-chip vacuum transistor based on field emission from carbon nanotubes via silicon micromachining processes was proposed by C. Bower et al. [11], achieving high frequency (10 GHz) and low control voltage (50–100 V) operation but still suffering from complex fabrication processes and inadequate emission stability and reliability of the carbon nanotubes emitters.

Intense research over the past decades has mainly focused on field emission vacuum transistors. Nevertheless, an intense local electric field with a typical magnitude of several volts per nanometer is required to induce field emission, which makes sharp tips preferred for field emitters with a large field enhancement factor and field emission quite sensitive to the microstructures of emitters. In addition, the electrons can ionize ambient gases such as oxygen and nitrogen with an enough high kinetic energy. Being accelerated along the electric field due to the cathode voltage, the positive ions will collide with the emitters and cause mechanical degradation of emitters leading to an unstable electron emission [8]. Moreover, field emission is quite sensitive to the absorption of ambient molecules onto emitters, which can change the work function of the emitters [18]. Therefore, an ultrahigh vacuum is required for stable field emission. As a result, field emission vacuum transistors still encounter the problems of unstable electron emission, poor uniformity and repeatability, and complex processing. Compared to field emission, thermionic electron emission is much less sensitive to the microstructures of the emitter and molecule absorption, indicating much more controllable and stable electron emission. A few years ago, a graphene-based vacuum transistor (GVT) was demonstrated by employing an electrically biased graphene as the thermionic electron emitter [19]. The GVT exhibits promising performances in several aspects, such as high ON/OFF ratio, small subthreshold slope, and low operating voltages. Importantly, this device provides a feasible way to achieving vacuum channel transistors based on thermionic electron emission. However, the GVT shows disadvantages of large leakage current and a high-power consumption.

In this paper, we report the scaling down of vacuum triodes to the microscale on a chip by employing a single Joule-heated multiwalled carbon nanotube (CNT) as the filament for thermionic electron emission. The CNT-based vacuum transistor (CVT) can be switched by tuning the bias voltage applied to the heavily doped silicon substrate (bottom gate) with an ON/OFF current ratio up to 10^4, and a subthreshold slope of ~4 $\text{V}\cdot\text{dec}^{-1}$. We also study the dependence of gate controllability on the distance between collector electrodes and emitter. The simulation of the electric field at the surface of CNTs indicates that the electric field in our devices is about one order of magnitude smaller than that in the devices based on field electron emission.

2. Materials and Methods

The schematic structure of a CVT is shown in Figure 1a, where a multiwalled CNT, acting as a filament, is freely suspended above a heavily doped Si substrate and pressed between Au/Cr electrodes. To excite thermionic electron emission from the CNT, a bias voltage (V_{Driven}) is applied to the suspended CNT, thus an electrical current (I_D) will pass through and heat it by self-Joule heating. A pair of Au/Cr electrodes with a collecting voltage (V_C) of 50 V are beside the CNT acting as the collector electrodes. As demonstrated in pervious works, thermionic electrons emit from a Joule-heated carbon nanotube while a bias voltage applied to it is larger than a threshold value, and the emission current increases exponentially with the bias voltage [20–22]. To switch emission current collected by the collector electrodes (I_C), the heavily doped Si substrate underneath the suspended CNT works as the gate electrode with a bias voltage (V_G) of −10–40 V. Figure 1b shows a scanning electron microscope (SEM) image of a CVT, and the magnified SEM image of the center area of the CVT is shown in Figure 1c, in which the distance between collector electrodes and electron emitter is approximately 1.5 μm.

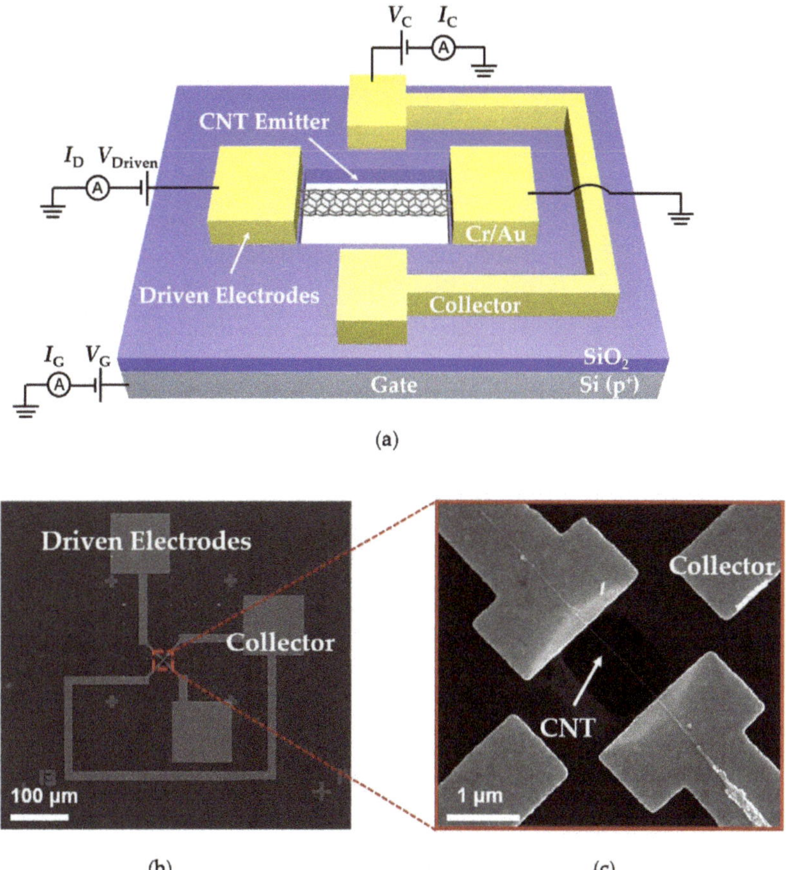

Figure 1. (**a**) A schematic illustration and working principles of a CNT-based vacuum transistor; (**b**) an SEM image of a global CVT; (**c**) an SEM image of the framed area in image (**b**).

CNT-based vacuum transistors are fabricated on SiO_2/Si wafer substrates by microfabrication technologies, where the SiO_2 layer thickness is 300 nm. The metallic multiwalled carbon nanotubes fabricated by arc discharge with a smooth surface and a perfect structure are dispersed on the substrate in alcohol solution. A proper carbon nanotube is selected first for device fabrication via SEM observation. The average diameter of selected carbon nanotubes is about 10 nm. Next, a pair of Au/Cr (80 nm/10 nm) driven electrodes and collector electrodes are fabricated by electron-beam lithography (EBL), electron-beam evaporator deposition, and a standard lift-off process, successively. To enhance the electron emission from a Joule-heated CNT [23,24], the SiO_2 layer underneath CNT must be removed by chemical etching to make the multiwalled CNT suspended over the Si substrate. A layer of polymethyl methacrylate (PMMA) is first used as a mask and the area for etching is defined by EBL. Second, the sample is plunged into buffered hydrofluoric acid for 330 s to remove the SiO_2 layer. After washing off the PMMA mask in acetone and drying the sample in hot isopropanol, a CNT-based vacuum transistor as shown in Figure 1 is finally obtained.

3. Results and Discussion

3.1. Thermionic Emission from a Multiwalled Carbon Nanotube

The performances of CVTs are measured on a probe station at room temperature with the vacuum level of about 1×10^{-2} Pa utilizing a Keithley 4200 semiconductor characterization system. The emission current of a multiwalled CNT electron emitter is measured repeatedly with collector electrodes applied with 50 V and gate electrode vacant. The device exhibits repeatable electron emission for five different measurements, as shown in Figure 2a. There is no emission current collected by collector electrode until V_{Driven} is larger than ~3.6 V, and the maximum emission current can reach up to 10 nA when V_{Driven} is ~4.4 V. It can be seen from Figure 2a that I_{C} increases exponentially with V_{Driven}, in good agreement with previous observation of thermionic emission from a suspended CNT [20–22]. As the collecting voltage is fixed at 50 V and thus electric field at the surface of the CNT is almost unchanged during the measurements, electron emission is not governed by electric field. The electron emission from a Joule-heated CNT is attributed to thermionic emission mechanism with non-thermal equilibrium electron distribution [20,21].

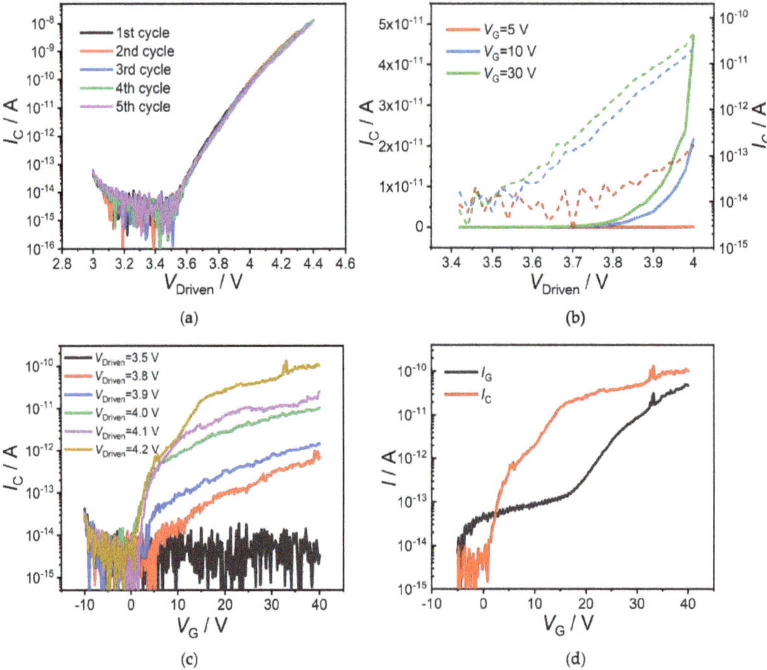

Figure 2. The performances of CVTs with distance between collector electrode and CNT emitter of 1.5 μm measured at room temperature and ~1×10^{-2} Pa. (**a**) The emission performance of a CNT emitter is measured repeatedly for five times when V_{C} is fixed at 50 V; (**b**) output characteristic ($I_{\mathrm{C}}-V_{\mathrm{Driven}}$ curves) and (**c**) transfer characteristic ($I_{\mathrm{C}}-V_{\mathrm{G}}$ curves) of a CVT at various gate voltages and driven voltages. The dashed lines and solid lines in (**b**) are in exponential and linear scale, respectively; (**d**) transfer characteristic of the same device in (**c**) when V_{C} is 50 V and V_{Driven} is 4.2 V, which indicates an ON/OFF current ratio up to 10^4. The black line shows the relationship between I_{G} and V_{G}.

3.2. Output Characteristic and Transfer Characteristic of a CNT-Based Vacuum Transistor

Figure 2b shows the output characteristic ($I_C - V_{Driven}$ curves) of a CVT at different gate voltages of 5 V, 10 V, 30 V, respectively. It can be seen from the curves plotted in exponential scale that the emission current increases exponentially with V_{Driven} and gate voltage also has a significant influence on emission current, implying a gate controllability. The transfer characteristic ($I_C - V_G$ curves) of a CVT with distance between collector electrode and emitter (D) of 1.5 μm under different driven voltages (V_{Driven}) applied to the multiwalled CNT is shown in Figure 2c when the collecting voltage is fixed at 50 V. It can be seen from the group curves that, when the gate voltage (V_G) is lower than a threshold voltage of ~0 V, the emission current is at a noise level and the CVT is in the OFF state. When V_G is larger than the threshold value, there is an obvious electron emission from the CNT emitter, indicating the CVT is switched to the ON state. The $I_C - V_G$ curves plotted in exponential scale demonstrate that the I_C increases exponentially with V_G above the threshold voltage. As is shown in the transfer characteristic curves, the threshold voltage decreases to a slight extent with the increase of V_{Driven}.

The bottom gate about 300 nm underneath the suspended CNT with a bias voltage enhances the strength of electric field around the CNT, thus lowering the surface barrier of the CNT, which enhances thermionic electron emission from the CNT. Comparing to pure thermionic emission, field-assisted thermionic emission can be well controlled by tuning electric field [25]. Therefore, we can control the states of a CVT through tuning the gate voltage. It can be seen obviously from Figure 2d that, the emission current increases fast with gate voltage at the low voltage regime (space charge regime), and then increases slowly with the gate voltage at high voltage regime (accelerating field regime). The gate controllability of CVTs is therefore attributed to the space charge effect and Schottky effect.

A single transfer characteristic curve together with a simultaneously measured gate current versus gate voltage curve is shown in Figure 2d. An ON/OFF current ratio as high as 10^4 and a subthreshold slope of ~4 $V \cdot dec^{-1}$ are observed. The subthreshold slope of the carbon nanotube vacuum transistors is not very satisfactory at present. The distance between the CNT emitter and gate electrode is ~300 nm. To maintain the same level of electric field, the shorter the distance, the lower the voltage required. In addition, the nanoscale curvature radius of CNTs surface induces a substantially larger local field enhancement. The field enhancement factor of a CNT will increase inversely with average diameter, and the field enhancement factor of a single-walled carbon nanotube can reach up to ~1404 [26]. On the other hand, the larger the coverage area of gate electrode to electron emitter, the stronger the gate controllability. The subthreshold slope of CVTs is not very satisfactory at present. In the future, limiting the distance between the electron emitter and gate electrode, using surround gate structure and decreasing the diameter of individual carbon nanotubes will be tried to improve the performance of the devices. It can be seen that the maximum leakage current collected by bottom gate is ~40 pA. This is approximately seven orders of magnitude smaller than that (~0.7 mA) of graphene-based vacuum transistor [16]. Importantly, the gate leakage current is much smaller than the collector current, which is important for a vacuum transistor.

3.3. Gate Controllability and Electric Field Strength Distribution

In order to optimize gate controllability, the dependence of gate controllability on the distance between collector electrodes and electron emitter is explored. Experimental data reveal that the distance (D) has an important influence on the bottom gate controllability. Figure 3a shows the transfer characteristic curves of CVTs with different distances of D = 0.5 μm and D = 1.5 μm, respectively, at V_C = 50 V and V_{Driven} = 4.0 V. When the distance is 1.5 μm, the emission current of the CVT increases rapidly with V_C in ON state and the ON/OFF current ratio is ~10^3. When the distance is 0.5 μm, however, the OFF current is so large that the slope of the $I_C - V_G$ curve is smaller than that of the former with a ON/OFF current ratio of ~10. Namely, the device with D of 0.5 μm is hard to be switched OFF. In addition, the emission current decreases slightly with the increase of V_G larger than

30 V, which is attributed to gate electrode-capturing electrons. Consequently, the longer the distance of collectors is, the higher performance in bottom gate controllability.

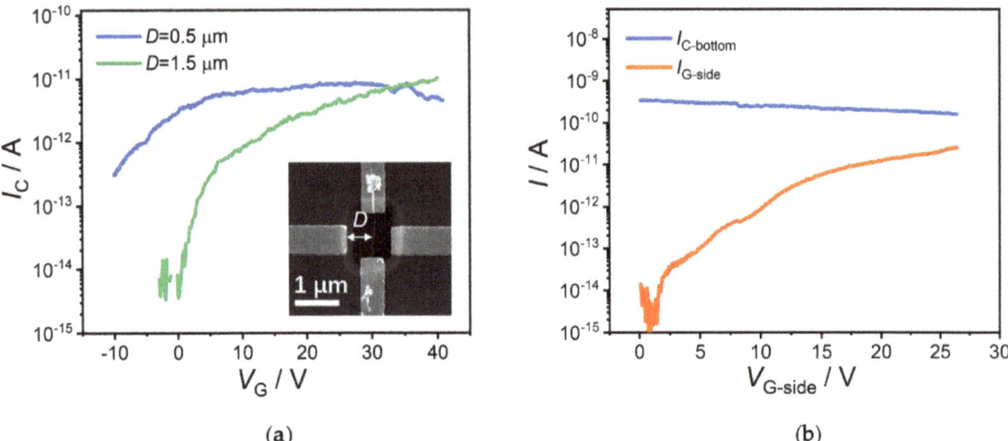

Figure 3. (**a**) Transfer characteristic of two CVTs with different distances between collector electrodes and CNT emitter: 0.5 μm and 1.5 μm. Inset: a scanning electron microscope image of a CVT with D = 0.5 μm; (**b**) simultaneously measured $I_C - V_G$ curve and $I_G - V_G$ curve of a CVT with the heavily doped Si substrate as the collector and side electrodes as the side-gate.

The transfer characteristic is also measured when the bottom Si substrate is set as the collector electrode (designated as bottom collector) and the side electrodes are set as the gate electrode (designated as side gate). Figure 3b shows the curve of bottom-collector current ($I_{C\text{-bottom}}$) versus side-gate voltage ($V_{G\text{-side}}$) curve and that of side-gate current ($I_{G\text{-side}}$) versus side-gate voltage ($V_{G\text{-side}}$) of a CVT. It can be clearly seen that the leakage current collected by side gate ($I_{G\text{-side}}$) increases rapidly with the side-gate voltage. However, the emission current collected by the bottom collector ($I_{C\text{-bottom}}$) maintains stable roughly with the side-gate voltage, indicating a weak gate controllability. The device configuration with the side electrodes as the collector and the bottom electrode as the gate therefore show much better gate controllability than the device configuration with the side electrodes as the gate and the bottom electrode as the collector.

To get further insights into the gate controllability of the devices, electric field in a CVT is calculated by COMSOL. The simulated CVT uses the same parameters as those of the CVT in experiments, including the thickness (90 nm) of metal electrodes, the thickness (300 nm) of SiO_2 layer, and the diameter (10 nm) and length (1 μm) of a CNT. To simulate the devices with different distance between the collector electrodes and CNT emitter, we calculated the electric field in the devices with D = 0.5 μm, 1 μm, and 1.5 μm, respectively. Figure 4a shows the distribution of electric field strength of a simulated device with D = 1.5 μm when V_G is fixed at 40 V. The distribution of electric field strength along the axis of a CNT emitter is shown in Figure 4b, where the strength of electric field at each axial position is obtained by averaging electric field strength along the circumference of the CNT. The distribution of electric field strength along the CNT axis can be well fitted by a curve of Gauss function. The maximum value of the fitting curve is regarded as the maximum electric field strength ($|E_{max}|$) in the CNT surface.

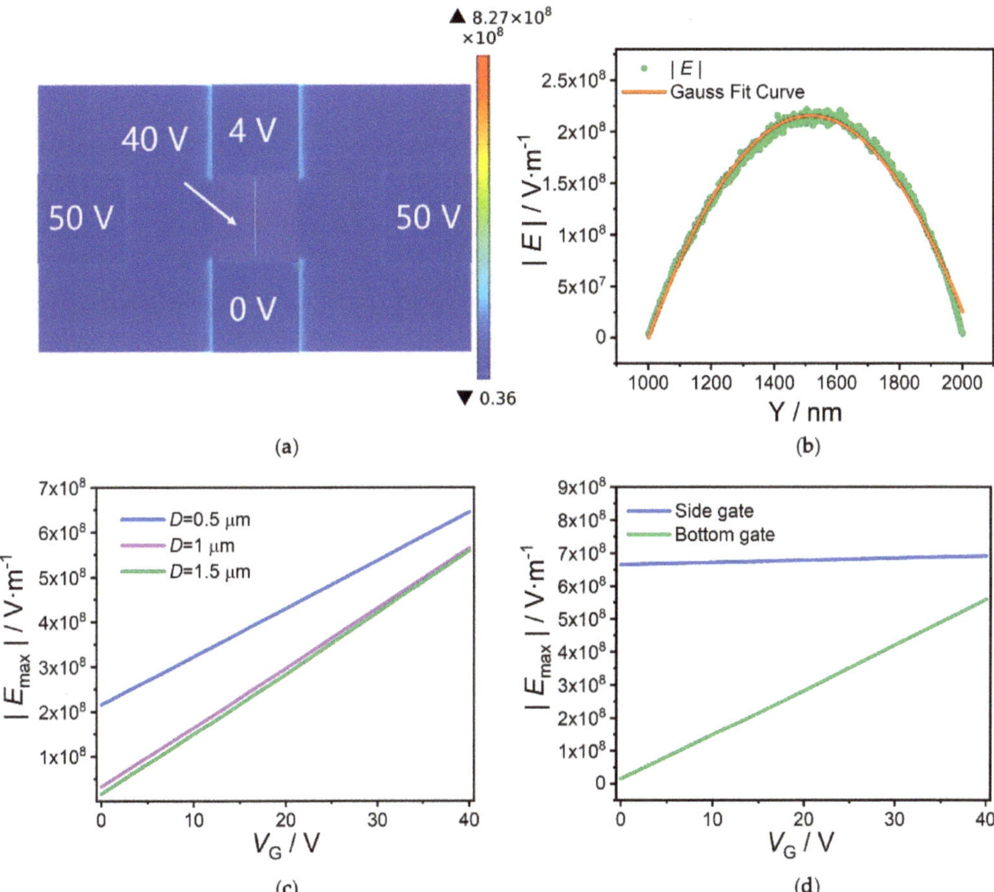

Figure 4. Electric field distribution simulations of a CNT-based vacuum transistor. (**a**) Electric field strength distribution in a simulation model with V_{Driven} = 4.0 V, V_C = 50 V, V_G = 40 V; (**b**) the distribution of electric field strength around the CNT surface along the axis of the carbon nanotube when D is 0.5 μm, V_{Driven} = 4.0 V, V_C = 50 V, V_G = 0 V; (**c**) $|E_{max}|-V_G$ curves for the devices with the side electrodes as the collector and D of 0.5 μm, 1 μm, 1.5 μm; (**d**) $|E_{max}|-V_G$ curves for the devices (D = 1.5 μm) with the bottom gate and side gate, respectively.

Figure 4c shows the dependence of $|E_{max}|$ on V_G for the devices with different D. It can be seen that $|E_{max}|$ increases from 1.63×10^7 V·m^{-1} to 5.60×10^8 V·m^{-1} by more than 30 times with the increase of V_G from 0 V to 40 V when D is 1.5 μm. In contrast, $|E_{max}|$ only increases by 2.99 and 17.26 times with the same increase of V_G from 0 V to 40 V corresponding to a D of 0.5 μm and 1.0 μm. The CVT with D of 1.5 μm therefore has a stronger gate controllability of $|E_{max}|$ by V_G than that of other CVTs with shorter D, which is in good agreement with the experimental results in Figure 3a. The poorer gate-to-emitter controllability for shorter collector-to-emitter distance can well be understood, by considering the fact that the electric field at the surface of CNT emitter is mainly governed by the collector but not the gate in the case of short collector-to-emitter distance.

We also calculate $|E_{max}|-V_G$ curves for different device configurations (the device with the side gate and bottom collector, and the device with the bottom gate and side collector). It can be seen from Figure 4d that, when the distance between CNT and side electrode is 1.5 μm and the distance between CNT and bottom electrode is 300 nm, the device with the side gate and bottom collector shows much less controllability of $|E_{max}|$ by V_G than the device with the bottom gate and side collector, in agreement with the experimental results in Figure 3b. The stronger gate controllability for device with the bottom gate is attributed to the dominance of electric field at CNT surface by the bottom electrode that is closer to the CNT emitter than the side electrodes.

Considering a gate voltage of 10 V for switching on a CVT, as shown in Figure 2c, an electric field with the maximum magnitude of as small as ~1×10^8 V·m^{-1} is needed to switch on a CVT according to Figure 4c. This is about one order of magnitude smaller than that needed for the vacuum transistors based on field emission, which requires a typical local threshold electric field of more than 1×10^9 V·m^{-1}. Compared with field emission that is quite sensitive to the microstructures of the emitter, field-assisted thermionic emission shows much less sensitivity to the microstructures of the emitter due to the much smaller threshold electric field [18]. Moreover, according to our previous works, the temperature of CNT emitter in a CVT can reach up to more than 2000 K [22]. The absorption of ambient molecules, which is thought to be the main reason for the instability of field emission, can be effectively prevented in such a hot electron emitter. The higher the temperature, the shorter the desorption time of the residual gas molecules. Our CVTs based on field-assisted thermionic emission are therefore expected to show better stability and uniformity than the vacuum transistors based on field emission.

4. Conclusions

In conclusion, we report a new structure of vacuum transistor based on field-assisted thermionic emission with individual CNTs as the filaments, fabricated by microfabrication technologies. The emission current can be controlled by tuning the bias voltage applied to the heavily doped Si substrate with an ON/OFF current ratio up to 10^4 and a subthreshold slope of ~4 V·dec^{-1}. Furthermore, we explore the influence of distance between collector electrodes and electron emitter on gate controllability and find that the longer the distance is, the stronger the gate controllability to electron emission is. Benefiting from field-assisted thermionic emission, the emission current of CVTs is much less sensitive to the microstructures of emitter and absorption of ambient molecules. The CVTs are therefore expected to show better stability and uniformity than the vacuum transistors based on field emission.

Author Contributions: Conceptualization, X.W. and Y.H.; methodology, Y.H. and Z.L.; software, Y.H. and S.M.; validation, Y.H.; formal analysis, Y.H., Z.L. and F.Z.; investigation, X.W. and Y.H.; resources, X.W.; data curation, Y.H.; writing—original draft preparation, Y.H.; writing—review and editing, X.W., Z.L. and Y.H.; funding acquisition, X.W. All authors have read and agreed to the published version of the manuscript.

Funding: This work was funded by the National Key Research and Development Program of China (Grant No. 2017YFA0205003, 2019YFA0210201) and the National Natural Science Foundation of China (Grant No. 62022007, 11874068, 11890671).

Data Availability Statement: The data presented in this study are available on request from the corresponding author. The data are not publicly available due to privacy.

Conflicts of Interest: The authors declare no conflict of interest.

References

1. Koomey, J.; Berard, S.; Sanchez, M.; Wong, H. Implications of Historical Trends in the Electrical Efficiency of Computing. *IEEE Ann. Hist. Comput.* **2011**, *33*, 46–54. [CrossRef]
2. Kim, H.K. Vacuum transistors for space travel. *Nat. Electron.* **2019**, *2*, 374–375. [CrossRef]
3. Brinkman, W.F.; Haggan, D.E.; Troutman, W.W. A history of the invention of the transistor and where it will lead us. *IEEE J. Solid-State Circuits* **1997**, *32*, 1858–1865. [CrossRef]
4. Barbour, E. The cool sound of tubes [vacuum tube musical applications]. *IEEE Spectr.* **1998**, *35*, 24–35. [CrossRef]
5. Stoner, B.R.; Glass, J.T. Nothing is like a vacuum. *Nat. Nanotechnol.* **2012**, *7*, 485–487. [CrossRef]
6. Liu, W.; Liu, Y.; Jia, Q.; Sun, B.; Chen, J. Terahertz laser diode using field emitter arrays. *Phys. Rev. B* **2021**, *103*, 035109. [CrossRef]

7. Han, J.-W.; Seol, M.-L.; Moon, D.-I.; Hunter, G.; Meyyappan, M. Nanoscale vacuum channel transistors fabricated on silicon carbide wafers. *Nat. Electron.* **2019**, *2*, 405–411. [CrossRef]
8. Nirantar, S.; Ahmed, T.; Ren, G.; Gutruf, P.; Xu, C.; Bhaskaran, M.; Walia, S.; Sriram, S. Metal–Air Transistors: Semiconductor-Free Field-Emission Air-Channel Nanoelectronics. *Nano Lett.* **2018**, *18*, 7478–7484. [CrossRef]
9. Jennings, S.G. The mean free path in air. *J. Aerosol Sci.* **1988**, *19*, 159–166. [CrossRef]
10. Han, J.-W.; Moon, D.-I.; Meyyappan, M. Nanoscale Vacuum Channel Transistor. *Nano Lett.* **2017**, *17*, 2146–2151. [CrossRef]
11. Bower, C.; Zhu, W.; Shalom, D.; Lopez, D.; Chen, L.H.; Gammel, P.L.; Jin, S. On-chip vacuum microtriode using carbon nanotube field emitters. *Appl. Phys. Lett.* **2002**, *80*, 3820–3822. [CrossRef]
12. Bhattacharya, R.; Han, J.W.; Browning, J.; Meyyappan, M. Complementary Vacuum Field Emission Transistor. *IEEE Trans. Electron Devices* **2021**, *68*, 5244–5249. [CrossRef]
13. Han, J.-W.; Seol, M.-L.; Kim, J.; Meyyappan, M. Nanoscale Complementary Vacuum Field Emission Transistor. *ACS Appl. Nano Mater.* **2020**, *3*, 11481–11488. [CrossRef]
14. Wang, X.; Xue, T.; Shen, Z.; Long, M.; Wu, S. Analysis of the electron emission characteristics and working mechanism of a planar bottom gate vacuum field emission triode with a nanoscale channel. *Nanoscale* **2021**, *13*, 14363–14370. [CrossRef]
15. Chang, W.T.; Pao, P.H. Field Electrons Intercepted by Coplanar Gates in Nanoscale Air Channel. *IEEE Trans. Electron Devices* **2019**, *66*, 3961–3966. [CrossRef]
16. Khoshkbijari, F.K.; Sharifi, M.J. Finger Gate Vacuum Channel Field Emission Transistors: Performance and Sensitivity Analysis. *IEEE Trans. Electron Devices* **2021**, *68*, 5250–5256. [CrossRef]
17. Fan, L.; Bi, J.; Xi, K.; Zhao, B.; Yang, X.; Xu, Y. Sub-10-nm Air Channel Field Emission Device with Ultra-Low Operating Voltage. *IEEE Electron Device Lett.* **2021**, *42*, 1390–1393. [CrossRef]
18. Terrones, M.; Terrones, H.; de Jonge, N.; Bonard, J.M. Carbon nanotube electron sources and applications. *Philos. Trans. R. Soc. Lond. Ser. A Math. Phys. Eng. Sci.* **2004**, *362*, 2239–2266.
19. Wu, G.; Wei, X.; Zhang, Z.; Chen, Q.; Peng, L. A Graphene-Based Vacuum Transistor with a High ON/OFF Current Ratio. *Adv. Funct. Mater.* **2015**, *25*, 5972–5978. [CrossRef]
20. Wei, X.; Golberg, D.; Chen, Q.; Bando, Y.; Peng, L. Phonon-Assisted Electron Emission from Individual Carbon Nanotubes. *Nano Lett.* **2011**, *11*, 734–739. [CrossRef]
21. Wei, X.; Wang, S.; Chen, Q.; Peng, L. Breakdown of Richardson's Law in Electron Emission from Individual Self-Joule-Heated Carbon Nanotubes. *Sci. Rep.* **2014**, *4*, 5102. [CrossRef] [PubMed]
22. Wang, Y.; Wu, G.; Xiang, L.; Xiao, M.; Li, Z.; Gao, S.; Chen, Q.; Wei, X. Single-walled carbon nanotube thermionic electron emitters with dense, efficient and reproducible electron emission. *Nanoscale* **2017**, *9*, 17814–17820. [CrossRef] [PubMed]
23. Lazzeri, M.; Piscanec, S.; Mauri, F.; Ferrari, A.C.; Robertson, J. Electron Transport and Hot Phonons in Carbon Nanotubes. *Phys. Rev. Lett.* **2005**, *95*, 236802. [CrossRef] [PubMed]
24. Pop, E.; Mann, D.; Cao, J.; Wang, Q.; Goodson, K.; Dai, H. Negative Differential Conductance and Hot Phonons in Suspended Nanotube Molecular Wires. *Phys. Rev. Lett.* **2005**, *95*, 155505. [CrossRef]
25. Herring, C.; Nichols, M.H. Thermionic Emission. *Rev. Mod. Phys.* **1949**, *21*, 185–270. [CrossRef]
26. Jin, F.; Liu, Y.; Day, C.M.; Little, S.A. Enhanced electron emission from functionalized carbon nanotubes with a barium strontium oxide coating produced by magnetron sputtering. *Carbon* **2007**, *45*, 587–593. [CrossRef]

MDPI

Article

Multiple Dielectric-Supported Ridge-Loaded Rhombus-Shaped Wideband Meander-Line Slow-Wave Structure for a V-Band TWT

Yuxin Wang, Yang Dong, Xiangbao Zhu, Jingyu Guo, Duo Xu, Shaomeng Wang * and Yubin Gong

National Key Laboratory of Science and Technology on Vacuum Electronics, University of Electronic Science and Technology of China, Chengdu 610054, China; 202022022426@std.uestc.edu.cn (Y.W.); 202111022405@std.uestc.edu.cn (Y.D.); 202022021934@std.uestc.edu.cn (X.Z.); 202111022436@std.uestc.edu.cn (J.G.); xuduo1234567@hotmail.com (D.X.); ybgong@uestc.edu.cn (Y.G.)

* Correspondence: wangsm@uestc.edu.cn; Tel.: +86-2883201538

Abstract: A multiple dielectric-supported ridge-loaded rhombus-shaped meander-line (MDSRL-RSML) slow-wave structure (SWS) is proposed for a V-band wideband traveling wave tube (TWT). The high-frequency and transmission characteristics of the SWS are investigated. The proposed structure can realize stable output via attenuator and special phase-velocity jumping. Particle-in-cell (PIC) results indicate that, for a 7 kV, 0.1 A sheet-beam, the average output power can reach 60 W at 60 GHz and a 3 dB bandwidth of 9 GHz, with the corresponding gain and electron efficiency of 30.8 dB and 17.2%, respectively. Compared with the dielectric-supported rhombus-shape meander-line (DS-RSML) SWS, the proposed structure has a wider bandwidth, higher gain, more stable structure, and better heat dissipation ability, which make it a good candidate source in millimeter-wave communications.

Keywords: meander-line; phase-velocity jumping; thermal analysis; slow-wave structure; traveling wave tube

Citation: Wang, Y.; Dong, Y.; Zhu, X.; Guo, J.; Xu, D.; Wang, S.; Gong, Y. Multiple Dielectric-Supported Ridge-Loaded Rhombus-Shaped Wideband Meander-Line Slow-Wave Structure for a V-Band TWT. *Electronics* **2022**, *11*, 405. https://doi.org/10.3390/electronics11030405

Academic Editor: Je-Hyeong Bahk

Received: 10 December 2021
Accepted: 26 January 2022
Published: 28 January 2022

1. Introduction

The traveling wave tube (TWT) is an important type of vacuum electronic device, which is widely used in radar, satellite communication, medical imaging electronic countermeasures, and other fields because of its excellent characteristics of high power, wide band, low noise, and high gain [1]. The traditional TWTs mainly include helix TWTs and coupled cavity TWTs. However, when the operation frequency is increased to the millimeter-wave or terahertz band, the traditional helix traveling wave tube faces great challenges during fabrication due to its small size [2]. At this time, there is an urgent demand for high-efficiency and high-power TWTs, and the planar TWT which is compatible with higher-precision microelectronics machine system (MEMS) technology has become a new research hotpot [3–5]. The planar TWT mainly refers to planar slow-wave structures (SWSs), including rectangular spiral SWSs [6–9], microstrip meander-line SWSs [10–12], staggered double-gate SWSs [13,14], and folded waveguide (FWG) SWSs [15].

Microstrip meander-line SWSs have been studied intensively for years, because of their advantages of easy integration, easy processing, and suitability for sheet-beam operation [16–19]. For instance, a V-band microstrip meander-line SWS, which was proposed in [19], could achieve rapid production using magnetron sputtering and laser ablation micromachining techniques. However, this kind of traditional microstrip meander-line usually uses a dielectric substrate to support the SWS, which is not resistant to electron bombardment and has the problem of charge accumulation and short-circuit.

In order to solve these problems, a new type of meander-line supported by dielectric rods was proposed in [20–23], in which the dielectric substrate was replaced by side rods. The so-called dielectric-rod-supported metallic meander-line SWS is suitable for a dual-sheet electron beam, thus having large power capability. For example, the Ka-band meander-line SWS supported by dielectric rods mentioned in [21] can achieve a gain of 23.4 dB at a voltage of 10.6 kV and a bandwidth of 6 GHz; the Ka-band ring-bar SWS supported by dielectric rods proposed in [22] can achieve a gain of 22.6 dB at 9.7 kV and a bandwidth of 2 GHz.

A dual-beam rhombus-shaped meander-line (DS-RSML) SWS supported from both sides by dielectric rods was proposed in [24], as shown in Figure 1a. This SWS can operate at a higher frequency with a wider lateral dimension compared to the dielectric-rod-supported metallic meander-line SWS. The particle-in-cell simulation showed that it can reach 83 W average output power at 58 GHz with a sheet electron beam of 7.7 kV, 0.1 A. However, its 3 dB bandwidth was only 6.5 GHz, and it struggled with heat dissipation. It was found that the ridge loading method can effectively broaden the bandwidth and reduce the operating voltage of the microstrip meander line [25]. Therefore, a multiple dielectric-supported ridge-loaded rhombus-shaped meander-line (MDSRL-RSML) SWS is proposed in this paper, as shown in Figure 1b.

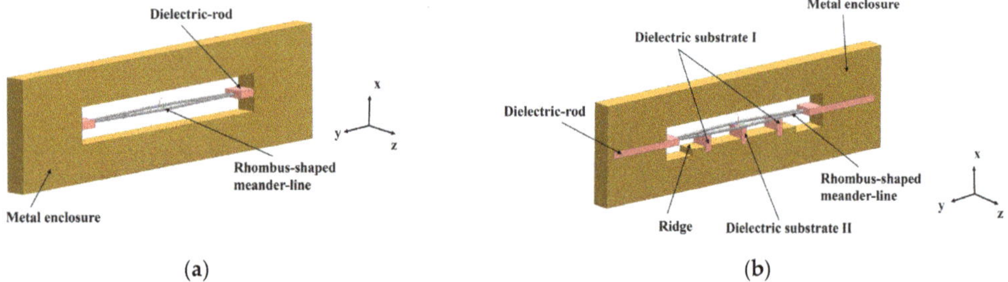

Figure 1. Single period of (**a**) DS-RSML SWS and (**b**) MDSRL-RSML SWS models.

The proposed structure of the DS-RSML SWS weakens the dispersion intensity of the TWT by employing metal ridges on the inner surface of the metal shell, so as to reduce the operation voltage and improve the working bandwidth of the TWT. Meanwhile, multiple dielectric rods embedded in the metal shell and ridge are applied on both sides and the bottom of the meander line, effectively solving the heat dissipation problem of the SWS and making the structure more stable. In addition, upon increasing the thickness of the dielectric substrate I along the propagation direction, the structural phase velocity is changed, which suppresses the backward wave oscillation and reduces the design difficulty. A glossary of abbreviations in the article is presented in Table 1.

Table 1. Glossary of abbreviations.

Abbreviations	The Full Name
SWS	Slow-wave structure
TWT	Traveling wave tube
MDSRL-RSML	Multiple dielectric-supported ridge-loaded rhombus-shaped meander-line
DS-RSML	Dielectric-supported rhombus-shape meander line
PIC	Particle-in-cell
MEMS	Microelectronics machine system
FWG	Folded waveguide
ML	Meander-line

The structure design and settings of the MDSRL-RSML, including the eigenmode structure, input and output coupling structure, and thermal analysis structure, are described in Section 2. Section 3 presents the high-frequency characteristics, cold and hot simulation results, and thermal analysis results. In Section 4, the results are analyzed and discussed. A short conclusion is presented in Section 5.

2. Materials and Structures

2.1. Single-Period Model of the MDSRL-RSML TWT

As shown in Figure 1b, the proposed structure consists of a metal enclosure with a ridge on the bottom edge, a rhombus-shaped meander-line, a pair of side supporting rods, and three bottom supporting dielectric rods. Taking processing into consideration, it is easier to achieve the clamping between the dielectric rods and the meander-line, as well as the connection between different periods of the meander-line, using a circular arc or straight line instead of inner and outer corners.

In the simulation, the material of the ridge-loaded metal enclosure and the rhombus-shaped meander-line is copper with a conductivity of 3×10^7 S/m, which was achieved by magnetron sputtering on a molybdenum surface (conductivity of 2×10^7 S/m) with a 3 μm thick copper layer (conductivity of 5.8×10^7 S/m) [26]. The dielectric rods are made from boron nitride (BN, relative dielectric constant of 5.12 and loss tangent of 0.005). The side supporting dielectric rods mainly play the role of clamping and stabilization. The bottom supporting dielectric rods are fixed in the ridge, supporting the meander-line from bottom up; they are used for support, phase-velocity taper, and heat dissipation.

The configuration of a single period of the proposed SWS with labeled dimensional parameters is presented in Figure 2a, and the specific shape of the rhombus-shaped meander-line is given in Figure 2b.

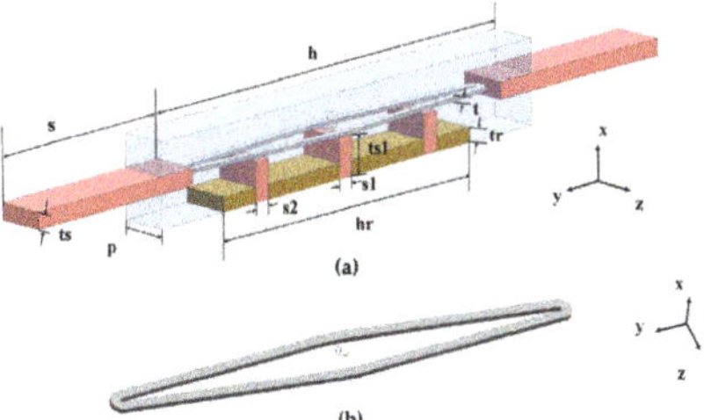

Figure 2. (**a**) Dimensional parameters of the MDSRL-RSML SWS; (**b**) rhombus-shaped meander-line.

The periodic length and the transverse length of the meander-line are represented by p and h, respectively. Table 2 shows the main dimensional parameters of the MDSRL-RSML SWS.

Table 2. Optimized dimensional parameters of the MDSRL-RSML SWS.

Parameters	Value (mm)	Parameters	Value (mm)
h	2	s	1
t	0.02	s1	0.078
p	0.23	s2	0.08
hr	1.6	ts	0.08
tr	0.07	ts1	0.19

2.2. Complete Transmission and Interaction Model of the MDSRL-RSML TWT

Figure 3a shows a schematic diagram of the overall MDSRL-RSML SWS. The entire SWS can be divided into five segments exclusive of the input and output parts, and each segment adopts different s2 parameters (the thickness of the dielectric rods I) to suppress the backward-wave oscillation of the structure instead of pitch tapering. In addition, the geometry of the attenuator is described in Figure 3b, where the two dielectric rods and dielectric rods II are coated with attenuating material for a 15-period length, while the material of the attenuator is beryllium oxide with a dielectric constant of 6.5 and loss tangent of 0.5. In Figure 3c, the structures at both ends of the slow-wave line include three periods of gradient with a height ratio of *R* to reduce reflection. Furthermore, in order to connect the meander-line SWS to a WR-15 rectangular waveguide port with reduced reflection, a stepped single-ridge waveguide is used, as shown in Figure 3d.

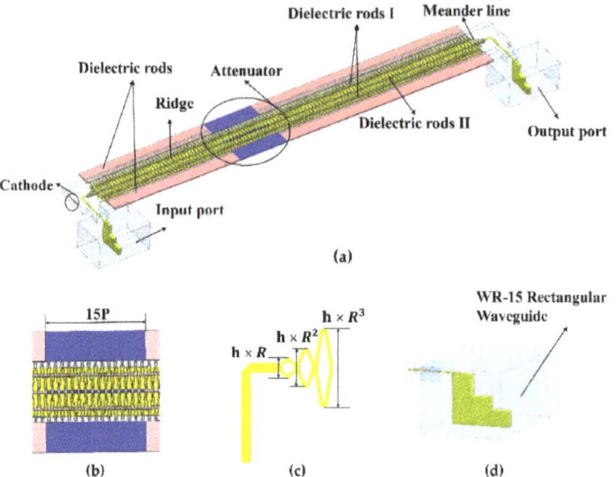

Figure 3. (**a**) The whole MDSRL-RSML SWS structure; (**b**) attenuator section; (**c**) gradient section; (**d**) the input–output couplers of the stepped ridge waveguide transition.

The optimized thickness of the dielectric rod I (s2) varies with the number of axial periods as shown in Figure 4.

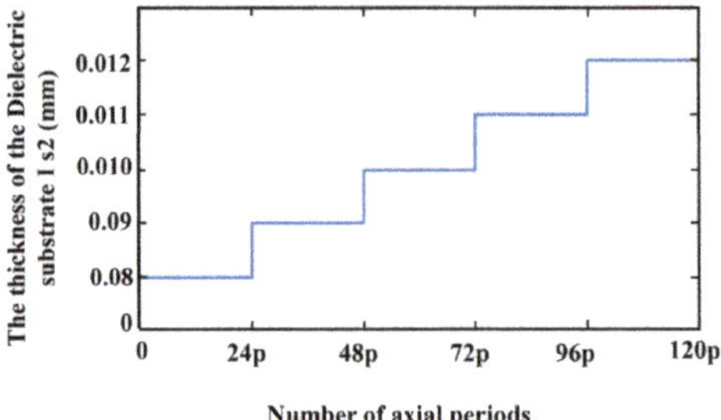

Figure 4. The thickness of the dielectric rod I (s2) as a function of the axial period.

2.3. Thermal Analysis Model of the MDSRL-RSML TWT

The tapered-S2 SWS with a length of 29.97 mm was used for the thermal analysis. The background temperature was 300 K, while the thermal conductivity of the material was as described in Table 3.

Table 3. Material properties in the simulation.

Material	Thermal Conductivity (W/K/m)
Copper	401
BN	60
BEO	250

In order to calculate thermal loss, we divided the whole SWS into 17 sections. The input and output parts were in one section each, with a length of 1.185 mm. The middle part contained 15 sections; the attenuator part had a length of 3.45 mm, 13 sections all had a length of 2.3 mm, and the last section had a length of 1.15 mm. The power flow curve in the PIC simulation was segmented to calculate the transmitted power of each part. Since S21 was −12.07 dB at 60 GHz without an attenuator, its section was about 0.4027 dB/mm. The power loss of each section could be calculated as follows [27]:

$$P_{loss} = \frac{P_{transmitted}}{10^{0.1 \times S_{21\_section}}} - P_{transmitted}. \quad (1)$$

The calculated power loss for each section is listed in Table 4 with a total loss of 35.34 W. Then, the power lost in each segment was set to be the heat source, and the shell was set to convection with a convective coefficient of 50 W/mm^2·K.

Table 4. Power loss for each section.

No.	1	2	3	4	5	6	7
P_{loss}(W)	0.0139	0.0241	0.0261	0.0491	0.1177	0.1855	0.2456
No.	8	9	10	11	12	13	14
P_{loss}(W)	0.7748	1.8581	3.9044	7.8145	13.4085	6.7522	0.1656

3. Simulation Results

3.1. High-Frequency Characteristics of the MDSRL-RSML TWT

The CST Eigenmode Solver was adopted to obtain the high-frequency characteristics of the MDSRL-RSML fundamental mode. For the dimension values in Table 2, the dispersion and coupling impedance curves of the DS-RSML and MDSRL-RSML are shown in Figure 5a,b. Although the former structure showed a higher coupling impedance in the low-frequency band, the latter structure had a lower normalized phase velocity and flatter dispersion characteristics, indicating a lower operating voltage and broader cold bandwidth, respectively.

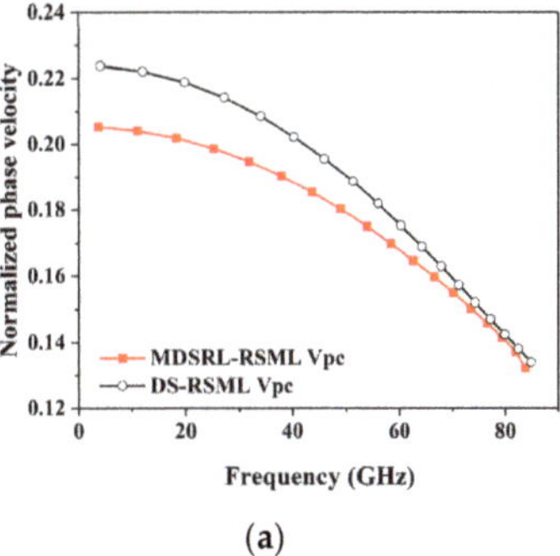

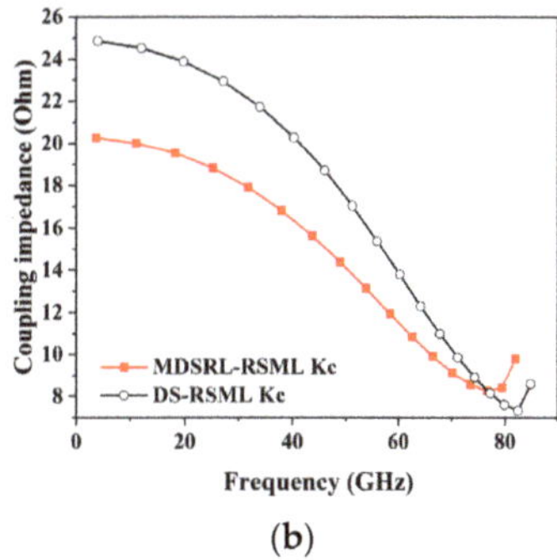

Figure 5. Comparison of (**a**) dispersion and (**b**) coupling impedance characteristics between DS-RSML and MDSRL–RSML SWS.

Figure 6 shows the electric field axial-component distributions of the fundamental mode and the higher-order mode along the transverse direction, both of which were strong in the middle and weak at both ends, indicating that the electric fields were suitable for the work of sheet-beam. However, this also suggests that, when the position of the electron beam was in the middle, the higher-order mode could compete with the fundamental mode.

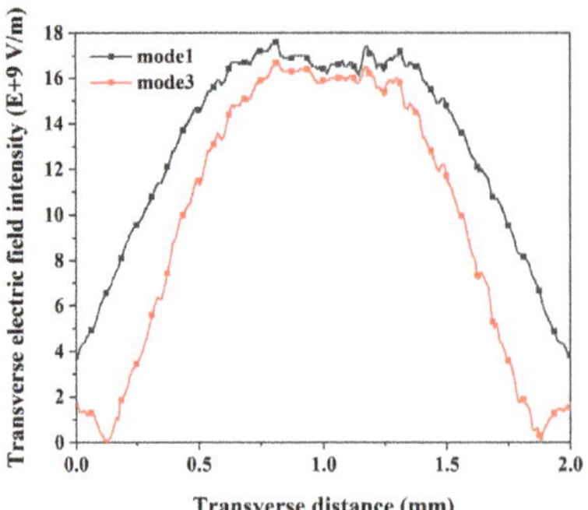

Figure 6. Transverse electric field distribution of fundamental mode and higher-order mode.

Figure 7a,b show the ω–β diagram and normalized phase velocity for the different values of s2, respectively. When s2 = 0.08 mm, the beam line, as shown in Figure 7a, intersected with each eigenmode at the following frequencies: mode 1 in the forward-wave region of 60 GHz, which could realize wave amplification; the intersection points with the higher-order modes 3 and 5 were located at 93 GHz and 175 GHz, respectively, in the backward-wave region, facilitating backward-wave oscillations.

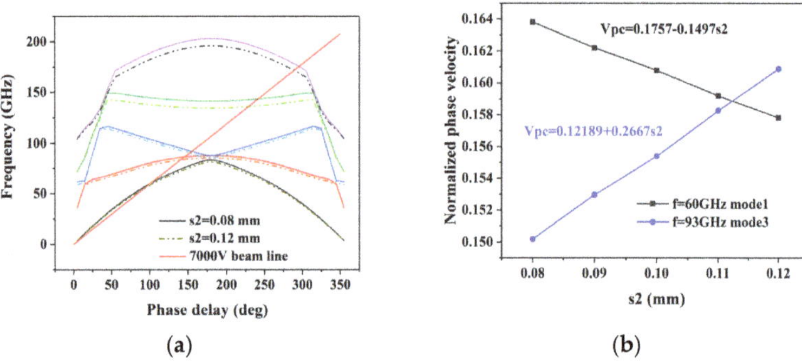

Figure 7. (**a**) ω–β diagram and (**b**) normalized phase-velocity characteristics with different s2 (in mm).

In Figure 7a, it can also be clearly seen that, as s2 increased, the eigenfrequency of the fundamental mode decreased much more slowly than the higher-order modes, which could effectively separate the operation voltages between the fundamental mode and higher-order modes and destroy their synchronization. Figure 7b gives a more intuitive description; as s2 changed, the normalized phase velocity of higher-order modes became more sensitive, and the corresponding voltage changed more than the fundamental mode. As a result, changing the rod width can be used as an effective method to suppress backward-wave oscillation.

3.2. Transmission Characteristics of the MDSRL-RSML TWT

The transmission characteristics of the 120-period MDSRL-RSML SWS together with the input–output couplers were determined using CST Microwave Studio [28]. The simulation results with and without attenuator are compared in Figure 8. Obviously, the attenuator only slightly changed S11, which was below −15 dB over the frequency range of 54–65 GHz, while S21 decreased about −15 dB across the whole frequency band, indicating that the attenuator effectively absorbed the wave energy.

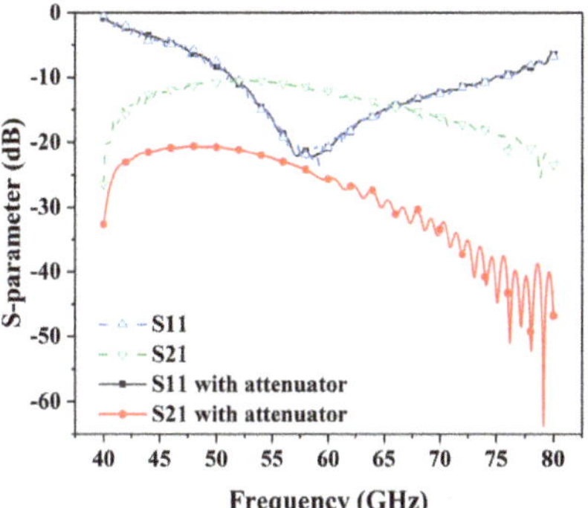

Figure 8. The S–parameters of the MDSRL-RSML SWS.

3.3. "Hot" Performance of the MDSRL-RSML TWT

In order to study the "hot" performance of the MDSRL-RSML SWS TWT, PIC simulations were conducted using CST Particle Studio [28]. An electron beam with a voltage of 7000 V and current of 0.1 A (current density of 143 A/cm^2) was employed, and the height and width of the sheet-beam were 1 mm and 0.07 mm, respectively. In addition, a solenoid focusing magnetic field of 0.6 T was used to maintain the beam shape, and the number of periods of the TWT was 120 with a whole length of 31 mm.

For an input signal of 0.05 W, the corresponding signal diagram and spectrum diagram at 60 GHz were obtained as shown in Figure 9a,b, respectively. Due to the combination of attenuator and phase-velocity jumping techniques, a stable output was achieved, and the signal was effectively amplified. Meanwhile, it can be seen from Figure 9b that the spectrum was pure; the harmonic components were about −50 dB smaller than the fundamental frequency, and the oscillation risk was effectively suppressed.

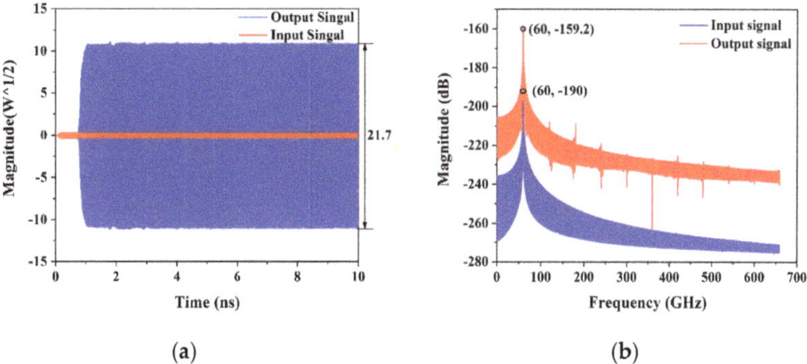

Figure 9. (**a**) Variation of input and output signals with time; (**b**) Fourier spectrum diagram of the output signal.

Figure 10 shows the electron beam bunching, as well as the phase space diagram of the electron beam, with most particles losing energy and a few electrons gaining energy, showing a sufficient beam wave energy transfer.

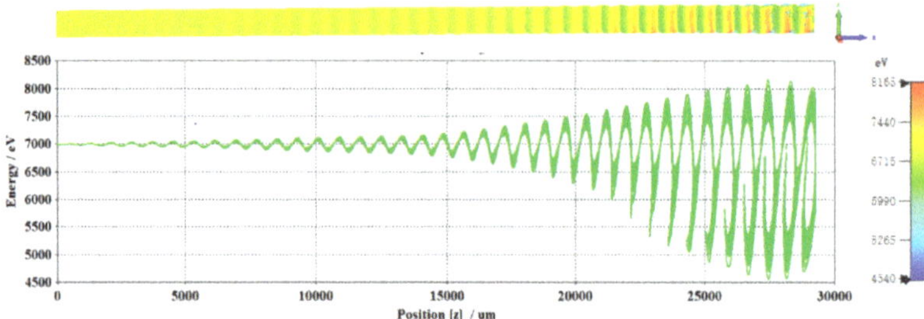

Figure 10. Phase space diagram.

The variation of average output powers of the DS-RSML and MDSRL-RSML at 60 GHz with the input power is depicted in Figure 11a,b. It can be seen that the maximum average output power of the former was 83 W at 0.17 W input power, with the corresponding saturation gain of 27 dB, while the saturation input power of the latter was reduced to 0.05 W, with the corresponding gain of 30.8 dB. Since the simulation was performed at the optimal operating voltages of both structures, the reduction in output power was to be expected. The former structure had an operating working voltage of 7700 V, while the optimal operating voltage of the latter structure was reduced to 7000 V.

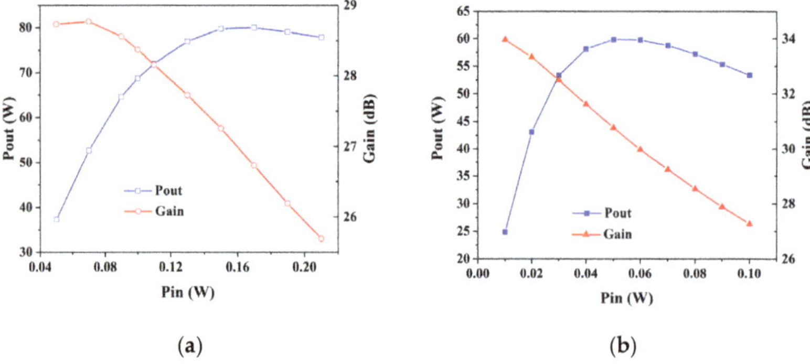

Figure 11. Variation of average output power and gain at 60 GHz with input power for the (a) DS-RSML SWS and (b) MDSRL-RSML SWS.

A comparison of DS-RSML and MDSRL-RSML in terms of the average output power and gain versus frequency is shown in Figure 12a,b. When the voltage was 7000 V, the maximum average output power of the MDSRL-RSML structure was 60 W at a frequency of 60 GHz, with an average gain of 30.8 dB and a 3 dB bandwidth of 9 GHz (53.5–62.5 GHz). The electron efficiency was 17.2%. The 3 dB bandwidth of MDSRL-RSML was about 1.5 times that of the DS-RSML; although the output power was reduced compared with the DS-RSML, it could achieve a higher gain.

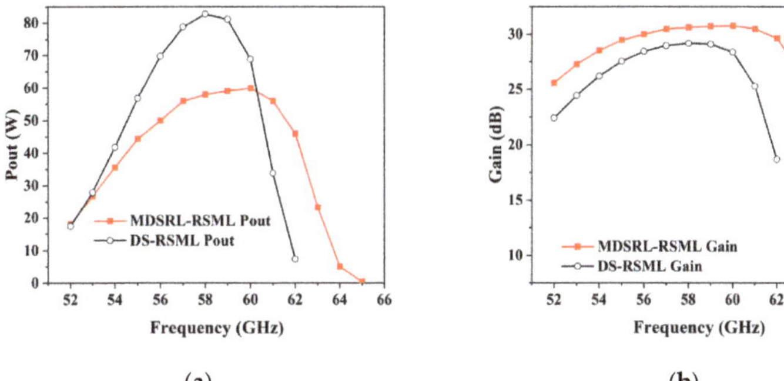

Figure 12. (**a**) Average output power and (**b**) gain versus frequency for the DS-RSML SWS and MDSRL-RSML SWS.

3.4. Thermal Distribution of the MDSRL-RSML TWT

For a traveling wave tube, good heat dissipation capability is a necessary condition for normal operation. To verify the heat-handling capacity of the proposed SWS, thermal analysis [29] of these two structures was performed using the CST thermal solver with the structure and settings described in Section 2.3. Here, we assumed that the flow rate of the electron beam was 100%, and only the heat loss due to ohmic loss was considered. Meanwhile, to simplify the thermal analysis, the thermal contact resistance (TCR) between different materials was not taken into consideration.

Figure 13 shows the temperature distribution of the two structures. In the DS-RSML, the maximum temperature distribution of each section was in the center of the meander line, because this section did not come into contact with the dielectric rods; thus, the maximum temperature could reach about 793.85 °C. On the other hand, the MDSRL-RSML had several supporting rods to help heat dissipation; thus, the maximum temperature could be maintained at 130.85 °C.

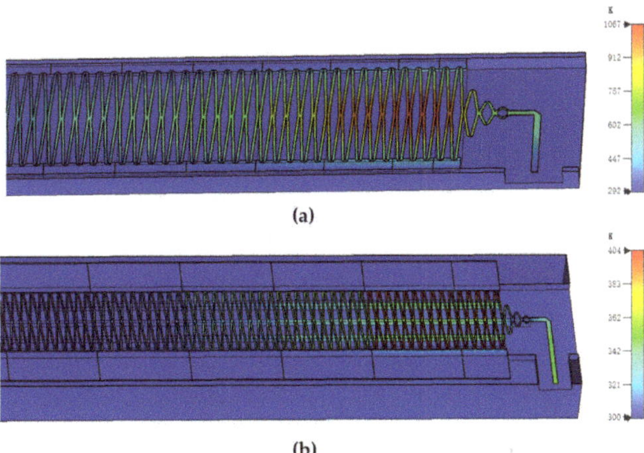

Figure 13. The maximum temperature distribution in the (**a**) DS-RSML SWS and (**b**) MDSRL-RSML SWS.

Figure 14 shows the temperature distribution along the axis of the DS-RSML SWS and MDSRL-RSML SWS. The proposed structure had better heat dissipation characteristics; accordingly, it would not suffer from excessively high temperature under a high-frequency power of tens of watts, effectively solving the heat dissipation problem in the previous structure.

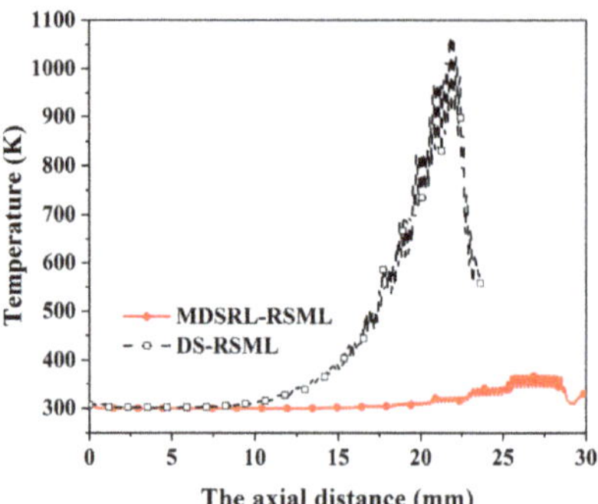

Figure 14. Temperature distribution along the axis of the DS-RSML SWS and MDSRL-RSML SWS.

4. Discussions

From the results presented in Section 3, it can be seen that the MDSRL-RSML SWS has the advantages of wide bandwidth and high gain. Furthermore, it offers a special method to suppress the backward-wave oscillation. Additionally, the problems of heat dissipation and stability of the SWS are solved in this structure.

Table 5 lists the key operation parameters of meander-line SWSs in the literature, as well as those of the proposed SWS; as it can be seen, compared with the specific parameters of the meander-line SWSs supported by dielectric rods [21,22,26], the gain of the proposed structure can exceed 30 dB while keeping a miniaturized structure. Compared with the structure in [18], the proposed structure has a lower operating voltage while maintaining a wide relative bandwidth. Compared with the structures in [19,23], the relative bandwidth of MSDRL-RSML was 2–3 times higher at a low operating voltage.

Table 5. Performance comparison of dielectric-supported meander-line SWSs.

Model	Frequency Band	Voltage (kV)	Gain (dB)	Bandwidth (GHz)	Relative Bandwidth
[21]	Ka	10.6	23.4	6	15.8%
[22]	Ka	9.7	22.6	2	5.3%
[26]	Ka	4.4	24.1	2.5	6.6%
MDSRL-RSML	V	7	30.8	9	15%

In addition, the main results presented in Section 3 were correlated using equations in MATLAB. For instance, the normalized phase velocity (*Vpc*) and s2 of mode 1 (V_{pc1}) and mode 3 (V_{pc3}) in Figure 7b approximately satisfied the following relationships:

$$V_{pc1} = 0.1757 - 0.1497s_2, \tag{2}$$

$$V_{pc3} = 0.12189 + 0.2667s_2. \tag{3}$$

As shown in Figure 11a, the relationship of output power and gain with input power of the DS-RSML structure could be approximately expressed as follows:

$$P_{out} = -23.84 + 1579P_{in} - 7727P_{in}^2 + 11970P_{in}^3, \tag{4}$$

$$Gain = 27.79 + 36.65P_{in} - 384.8P_{in}^2 + 776.4P_{in}^3. \tag{5}$$

As shown in Figure 11b, the relationship of output power and gain with input power of the MDSRL-RSML structure could be approximately expressed as follows:

$$P_{out} = -74.59 + 2958P_{in} - 22900P_{in}^2 + 73410P_{in}^3 - 83830P_{in}^4, \tag{6}$$

$$Gain = 34.33 + 329.2P_{in}^2 - 13180P_{in}^3 + 9180P_{in}^4 - 187700P_{in}^5. \tag{7}$$

5. Conclusions

An MDSRL-RSML SWS for V-band TWTs was proposed and investigated in this paper. The dispersion characteristics and coupling impedance were studied, revealing that the proposed structure has flatter dispersion characteristics and a lower beam voltage than DS-RSML. The proposed structure has good transmission characteristics, with S11 below −15 dB over the frequency range of 54–65 GHz. Stable output can be realized by adding an attenuator and using special phase-velocity jumping. The PIC result indicated a maximum average output power of 60 W at 60 GHz for a beam voltage of 7000 V and beam current of 0.1 A with a bandwidth of 9 GHz, with a corresponding gain and electron efficiency of 30.8 dB and 17.2%, respectively. Compared with DS-RSML, the MDSRL-RSML has a wider bandwidth and higher gain; furthermore, this structure solves the heat dissipation problem and improves the stability of the whole structure.

Author Contributions: Conceptualization, Y.W.; data curation, X.Z.; project administration, S.W.; Supervision, Y.G.; Validation, Y.D.; Writing—original draft, J.G.; Writing—review & editing, D.X. All authors have read and agreed to the published version of the manuscript.

Funding: This work was supported by the National Natural Science Foundation of China (NSFC) (Grant Nos. 61921002, 92163204, 61988102) and the Laboratory Fund of the National Key Lab on Microwave Vacuum Devices (Grant No. 19ZS202807200303A).

Conflicts of Interest: The authors declare no conflict of interest.

References

1. Chong, C.K.; Menninger, W.L. Latest Advancements in High-Power Millimeter-Wave Helix TWTs. *IEEE Trans. Plasma Sci.* **2010**, *38*, 1227–1238. [CrossRef]
2. Liu, L.; Wei, Y.; Shen, F.; Zhao, G.; Yue, L.; Duan, Z.; Wang, W.; Gong, Y.; Li, L.; Feng, J. A Novel Winding Microstrip Meander-Line Slow-Wave Structure for V-Band TWT. *IEEE Electron Device Lett.* **2013**, *34*, 1325–1327. [CrossRef]
3. Luo, J.; Feng, J.; Gong, Y. A Review of Microwave Vacuum Devices in China: Theory and Device Development Including High-Power Klystrons, Spaceborne TWTs, and Gyro-TWTs. *IEEE Microwave. Mag.* **2021**, *22*, 18–33. [CrossRef]
4. Paoloni, C.; Gamzina, D.; Letizia, R.; Zheng, Y.; Luhmann, N.C., Jr. Millimeter wave traveling wave tubes for the 21st Century. *J. Electromagn. Waves Appl.* **2021**, *35*, 567–603. [CrossRef]
5. Wang, Z.; Wang, H.; Xu, D.; He, T.; Li, X.; Gong, H.; Lu, Z.; Duan, Z.; Wei, Y.; Gong, Y. Study on Planar Slow Wave Structure of Traveling Wave Tube. *Vac. Electron.* **2018**, *2*, 20–30.
6. Chua, C.; Aditya, S. A 3-D U-Shaped Meander-Line Slow-Wave Structure for Traveling-Wave-Tube Applications. *IEEE Trans. Electron Devices* **2013**, *60*, 1251–1256. [CrossRef]
7. Chua, C.S.; Aditya, S.; Shen, Z.X. Planar Helix with Straight-Edge Connections in the Presence of Multilayer Dielectric Substrates. *IEEE Trans. Electron Devices* **2010**, *57*, 3451–3459. [CrossRef]
8. Chua, C.; Tsai, J.M.; Aditya, S.; Tang, M.; Ho, S.W.; Shen, Z.; Wang, L. Microfabrication and Characterization of W-Band Planar Helix Slow-Wave Structure with Straight-Edge Connections. *IEEE Trans. Electron Devices* **2011**, *58*, 4098–4105. [CrossRef]
9. Chua, C.S.; Tsai, M.L.J.; Tang, M.; Aditya, S.; Shen, Z.X. Microfabrication of a Planar Helix with Straight-Edge Connections Slow-Wave Structure. *Adv. Mater. Res.* **2011**, *254*, 17–20. [CrossRef]
10. Shen, F.; Wei, Y.; Xu, X.; Liu, Y.; Huang, M.; Tang, T.; Gong, Y. U-shaped microstrip meander-line slow-wave structure for Ka-band traveling-wave tube. In Proceedings of the 2012 International Conference on Microwave and Millimeter Wave Technology (ICMMT), Shenzhen, China, 5–8 May 2012.
11. Shen, F.; Wei, Y.; Yin, H.; Gong, Y.; Xu, X.; Wang, S.; Wang, W.; Feng, J. A Novel V-Shaped Microstrip Meander-Line Slow-Wave Structure for W-band MMPM. *IEEE Trans. Plasma Sci.* **2011**, *40*, 463–469. [CrossRef]
12. Shen, F.; Wei, Y.; Xu, X.; Lai, J.; Huang, M.; Zhao, G.; Gong, Y. Rhombus-shaped Microstrip Meander-line Slow-wave Structure for 140GHz Traveling-wave Tube. In Proceedings of the 2012 IEEE International Vacuum Electronics Conference (IVEC), Monterey, CA, USA, 24–26 April 2012.
13. Lu, Z.; Ge, W.; Wen, R.; Wang, Z.; Gong, H.; Wei, Y.; Gong, Y. 0.2-THz Traveling Wave Tube Based on the Sheet Beam and a Novel Staggered Double Corrugated Waveguide. *IEEE Trans. Plasma Sci.* **2020**, *48*, 3229–3237. [CrossRef]
14. Shao, W.; Tian, H.; Wang, Z.; Lu, Z.; Gong, H.; Tang, T.; Duan, Z.; Wei, Y.; Gong, Y.; Feng, J. Study for 850 GHz Sheet Beam Staggered Double-Vane Traveling Wave Tube Considering the Metal Loss. In Proceedings of the 2018 IEEE International Vacuum Electronics Conference (IVEC), Monterey, CA, USA, 21 April 2018.
15. Xu, D.; Shao, W.; He, T.; Wang, H.; Wang, Z.; Lu, Z.; Gong, H.; Duan, Z.; Feng, J.; Gong, Y. Investigation on 0.1 THz Array Beams Folded Waveguide Traveling Wave Tube. In Proceedings of the 2019 Photonics & Electromagnetics Research Symposium (Piers), Xiamen, China, 17–20 December 2019.

16. Ding, C.; Wei, Y.; Wang, Y.; Xu, J.; Tang, T.; Huang, M.; Zhang, L.; Li, Q.; Xia, L.; Gong, Y.; et al. 2-Dimensional Microstrip Meander-line for Broad Band Planar TWTs. In Proceedings of the 2016 IEEE International Vacuum Electronics Conference (IVEC), Monterey, CA, USA, 19–21 April 2016.
17. Wang, S.; Aditya, S.; Xia, X.; Ali, Z.; Miao, J. On-Wafer Microstrip Meander-Line Slow-Wave Structure at Ka-Band. *IEEE Trans. Electron Devices* **2018**, *65*, 2142–2148. [CrossRef]
18. Wang, S.; Aditya, S.; Xia, X.; Ali, Z.; Miao, J.; Zheng, Y. Ka-Band Symmetric V-Shaped Meander-Line Slow Wave Structure. *IEEE Trans. Plasma Sci.* **2019**, *47*, 4650–4657. [CrossRef]
19. Ryskin, N.M.; Rozhnev, A.G.; Starodubov, A.V.; Serdobintsev, A.A.; Pavlov, A.M.; Benedik, A.I.; Torgashov, R.A.; Torgashov, G.V.; Sinitsyn, N.I. Planar Microstrip Slow-Wave Structure for Low-Voltage V-band Traveling-Wave Tube with a Sheet Electron Beam. *IEEE Electron Device Lett.* **2018**, *1*, 5. [CrossRef]
20. He, T.; Li, X.; Wang, Z.; Wang, S.; Lu, Z.; Gong, H.; Duan, Z.; Feng, J.; Gong, Y. Design and Cold Test of Dual Beam Azimuthal Supported Angular Log-Periodic Strip-Line Slow Wave Structure. *J. Infrared Millimeter Terahertz Waves* **2020**, *41*, 785–795. [CrossRef]
21. Wang, H.; Wang, Z.; Li, X.; He, T.; Xu, D.; Gong, H.; Tang, T.; Duan, Z.; Wei, Y.; Gong, Y. Study of a miniaturized dual-beam TWT with planar dielectric-rods-support uniform metallic meander line. *Phys. Plasmas* **2018**, *25*, 63113. [CrossRef]
22. Dong, Y.; Chen, Z.; Li, X.; Wang, H.; Wang, Z.; Wang, S.; Lu, Z.; Gong, H.; Duan, Z.; Feng, J.; et al. Ka-band dual sheet beam traveling wave tube using supported planar ring-bar slow wave structure. *J. Electromagn. Waves Appl.* **2020**, *34*, 2236–2250. [CrossRef]
23. Wang, H.; Wang, S.; Wang, Z.; Li, X.; Xu, D.; Duan, Z.; Lu, Z.; Gong, H.; Aditya, S.; Gong, Y. Dielectric-Supported Staggered Dual Meander-Line Slow Wave Structure for an E-Band TWT. *IEEE Trans. Electron Devices* **2021**, *68*, 369–375. [CrossRef]
24. Wang, Y.; Dong, Y.; Wang, S.; Gong, Y. Dielectric-Supported Rhombus-Shaped Meander-Line Slow-Wave Structure for a V-band Dual-Sheet Beam Traveling Wave Tube. In Proceedings of the 2021 Photonics & Electromagnetics Research Symposium (Piers), Hangzhou, China, 21–25 December 2021.
25. Xu, D.; Wang, H.; He, T.; Li, X.; Lu, Z.; Gong, H.; Wang, Z.; Duan, Z.; Gong, Y. Study on Broadband Ridge-Loaded Symmetrical Conformal Microstrip Meander Line Traveling Wave Tube at Ka-Band. In Proceedings of the 2019 International Vacuum Electronics Conference (IVEC), Busan, Korea, 28 April 2019.
26. Wang, H.; Wang, S.; Wang, Z.; Li, X.; He, T.; Xu, D.; Duan, Z.; Lu, Z.; Gong, H.; Gong, Y. Study of an Attenuator Supporting Meander-Line Slow Wave Structure for Ka-Band TWT. *Electronics* **2021**, *10*, 2372. [CrossRef]
27. Zhao, C.; Aditya, S.; Wang, S.; Miao, J.; Xia, X. A Wideband Microfabricated Ka-Band Planar Helix Slow-Wave Structure. *IEEE Trans. Electron Devices* **2016**, *63*, 2900–2906. [CrossRef]
28. Introduction of CST Microwave Studio. Available online: https://www.cst.com/products/cstms (accessed on 1 December 2016).
29. Wang, S.; Zhao, C.; Aditya, S.; Chua, C. Thermal Characteristics of a Ka-Band Planar Helix Slow-Wave Structure. In Proceedings of the 2015 International Vacuum Electronics Conference (IVEC), Beijing, China, 27–29 April 2015.

Article

Broadband-Printed Traveling-Wave Tube Based on a Staggered Rings Microstrip Line Slow-Wave Structure

Ruichao Yang [1], Lingna Yue [1,*], Jin Xu [1,*], Pengcheng Yin [1], Jinjing Luo [1], Hexin Wang [1], Dongdong Jia [1], Jian Zhang [1], Hairong Yin [1], Jinchi Cai [1], Guo Guo [1], Guoqing Zhao [1], Wenxiang Wang [1], Dazhi Li [2] and Yanyu Wei [1]

1 National Key Laboratory of Science and Technology on Vacuum Electronics, University of Electronic Science and Technology of China, Chengdu 611731, China; ruichaoyang@foxmail.com (R.Y.); yyppcchh@163.com (P.Y.); jinjingluo_uestc@163.com (J.L.); whx427@126.com (H.W.); DongdJia@outlook.com (D.J.); jiann.zhang@outlook.com (J.Z.); hryin@uestc.edu.cn (H.Y.); jccai@uestc.edu.cn (J.C.); guoguo@uestc.edu.cn (G.G.); zhaogq@uestc.edu.cn (G.Z.); wxwang@uestc.edu.cn (W.W.); yywei@uestc.edu.cn (Y.W.)

2 Neubrex Company Limited, Kobe 6500023, Japan; dazhi_li@hotmail.com

* Correspondence: lnyue@uestc.edu.cn (L.Y.); alionxj@uestc.edu.cn (J.X.)

Abstract: To increase the output power of microstrip line traveling-wave tubes, a staggered rings microstrip line (SRML) slow-wave structure (SWS) based on a U-shaped mender line (U-shaped ML) SWS and a ring-shaped microstrip line (RML) SWS has been proposed in this paper. Compared with U-shaped ML SWS and RML SWS, SRML SWS has a wider transverse width, which means SRML SWS has a larger area for beam–wave interaction. The simulation results show that SRML SWS has a wider bandwidth than U-shaped ML SWS and a lower phase velocity than RML SWS. Input/output couplers, which consist of microstrip probes and transition sections, have been designed to transmit signals from a rectangular waveguide to the SWS; the simulation results present that the designed input/output structure has good transmission characteristics. Particle-in-cell (PIC) simulation results indicate that the SRML TWT has a maximum output of 322 W at 32.5 GHz under a beam voltage of 9.7 kV and a beam current of 380 mA, and the corresponding electronic efficiency is around 8.74%. The output power is over 100 W in the frequency range of 27 GHz to 38 GHz.

Keywords: microstrip line; slow-wave structure (SWS); traveling-wave tube (TWT); Ka-band

Citation: Yang, R.; Yue, L.; Xu, J.; Yin, P.; Luo, J.; Wang, H.; Jia, D.; Zhang, J.; Yin, H.; Cai, J.; et al. Broadband-Printed Traveling-Wave Tube Based on a Staggered Rings Microstrip Line Slow-Wave Structure. *Electronics* **2022**, *11*, 384. https://doi.org/10.3390/electronics11030384

Academic Editor: Geok Ing Ng

Received: 31 December 2021
Accepted: 19 January 2022
Published: 27 January 2022

1. Introduction

High average power at millimeter-wave frequency, which is lightweight, low voltage, compact, and broadband, is demanded in many significant applications such as electronic counter measures, radar, and communications [1]. Solid-state amplifiers are lightweight and compact, but the output power, bandwidth, and efficiency struggle to meet the requirements. Traveling-wave tubes (TWTs) such as helix TWT [2], coupled-cavity TWT (CC-TWT) [3], and folded-waveguide TWT (FW-TWT) [4] show great potential in this frequency range due to high-power output or broadband. However, the conventional traveling-wave tubes are heavy, high voltage, or hard to fabricate at millimeter-wave frequency.

Planar TWT has become an attractive interaction structure because it is lightweight, compact, and can be mass fabricated. Numerous investigations have been conducted on meander-line slow-wave structure (ML-SWS) [5–9] and its deformed structure [10–18] in recent years. As planar SWS is a 2D structure, the fabrication problem that conventional SWS such as helix, coupled-cavity, and folded-waveguide met when the operating frequency was Ka-band or above can be solved using micro-electromechanical systems (MEMS).

The conventional microstrip line SWSs have the advantage of miniaturization and low voltage, but these are also their flaws. This feature of miniaturization leads to the small electron beam current used by the microstrip TWT, which causes the output power of the microstrip TWT to be relatively low. As the force exerted by the high-frequency field on the sheet electron beam in the microstrip TWT is not symmetrical, and the beam voltage of the microstrip TWT is usually very low, focusing the sheet electron beam was a huge challenge.

To obtain a higher output power, a staggered rings microstrip line slow-wave structure (SRML SWS) based on a ring-shaped microstrip line slow-wave structure [18] (RML SWS) has been designed. By staggered placement of adjacent rings, the area for beam–wave interaction was increased, which resulted in an increased beam current with the same current density.

The content of this paper is arranged as follow. Firstly, the dispersion characteristics of SRML SWS are analyzed and discussed in Section 2. In Section 3, the transmission characteristics of SRML SWS with input/output couplers are presented. Section 4 shows the particle-in-cell (PIC) simulation results of the designed SRML TWT. Finally, the fabrication of SRML SWS and the results are discussed and concluded in Sections 5 and 6.

2. Dispersion Characteristics of SRML SWS

Figure 1a is the traditional U-shaped meander line slow-wave structure (ML SWS); by overlapping two U-shaped metal lines of the ML SWS with the opposite phases, a ring-shaped microstrip line slow-wave structure (RML SWS) [18], as shown in Figure 1b, was obtained. Additionally, as shown in Figure 1c, a staggered rings microstrip line slow-wave structure (SRML SWS) was constructed through periodically interlacing adjacent rings of the RML SWS. The SWS consisted of dielectric substrate and the metal line placed on the surface of the substrate. The staggered distance of the two rings is δ, the width of the ring is w_r, the period length is p, the inner gap of the ring is s, the width of the metal line is w, the thickness of the metal line is t, the width of the shielding box and substrate is a, the height of the shielding box is h_a, and the thickness of the dielectric substrate is h_d. The material of the dielectric substrate was quartz glass, and the metal line was copper made. The microstrip line SWS was placed in a metal-shielding cavity to maintain a vacuum state, and copper was the material used.

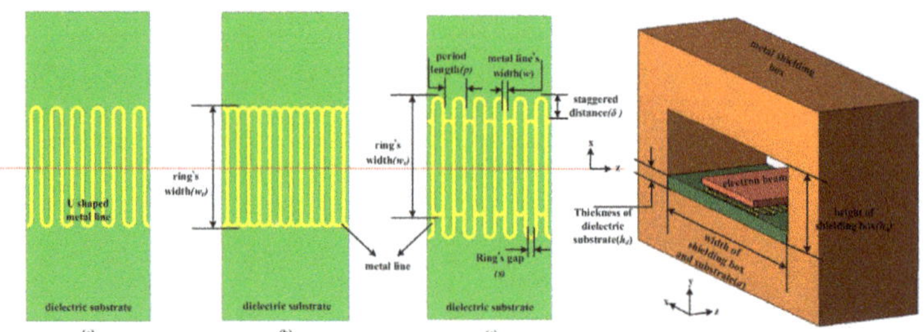

Figure 1. The diagram of (**a**) U-shaped ML SWS, (**b**) RML SWS, and (**c**) SRML SWS.

The phase velocity and interaction impedance are defined by the following:

$$v_p = \frac{\omega}{\beta_n} \tag{1}$$

$$K_{\text{cn}} = \frac{|E_{zn}|^2}{2\beta_n^2 P_\omega} \tag{2}$$

$$\beta_\text{n} = \beta_0 + \frac{2n\pi}{p} \qquad n = 0, \pm 1, \pm 2, \cdots \tag{3}$$

Here, ω is the angular frequency,P_ω is the transmission power along the longitudinal direction, E_{zn} is the amplitude of the longitudinal electrical field, p is the period length of the SWS and β_0 is the phase constant of fundamental wave, which equals ϕ/p, and ϕ is the phase shift in one period. By compiling and solving the above formulas in a field calculator of HFSS, the phase velocity and the interaction impedance of the SWS can be obtained.

Figure 2 presents the simulation results of the normalized phase velocity and the interaction impedance of the U-shaped ML SWS, RML SWS, and SRML SWS with the same ring width and period length. As shown in Figure 2a, the RML SWS and SRML SWS have larger bandwidths than that of the U-shaped ML SWS, but their phase velocity is far higher than that of the ML SWS. Additionally, by staggering the adjacent rings, the normalized phase velocity of the SRML SWS was significantly decreased when compared with the RML SWS. Additionally, the SRML SWS had a smaller change rate of normalized phase velocity, which means the SRML TWT had a wider working-frequency band. Figure 2b is a comparison diagram of the interaction impedance of the three structures. The interaction impedance was calculated 0.1 mm away from the upper face of the metal line. Figure 2b shows that the U-shaped ML SWS had a larger interaction impedance at the lower frequency band, but the interaction impedance of the SRML SWS and the RML SWS were larger at the Ka-band.

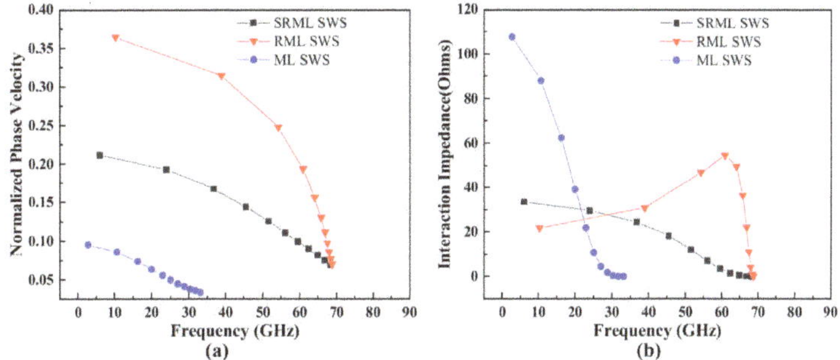

Figure 2. The comparisons between ML SWS, RML SWS, and SRML SWS. (**a**) Normalized phase velocity, (**b**) interaction impedance.

Additionally, Figure 3 shows that the interaction impedance varied with the position of the calculation line of coupling impedance at 32.5 GHz. It is obvious that the SRML SWS had a larger coupling area than the RML SWS, which means a wider electron beam can be used for the beam–wave interaction of SRML TWT. In other words, an SRML TWT can be driven by an electron beam with a higher beam current than that of an RML TWT at the same current density, and the SRML TWT would have a higher output power.

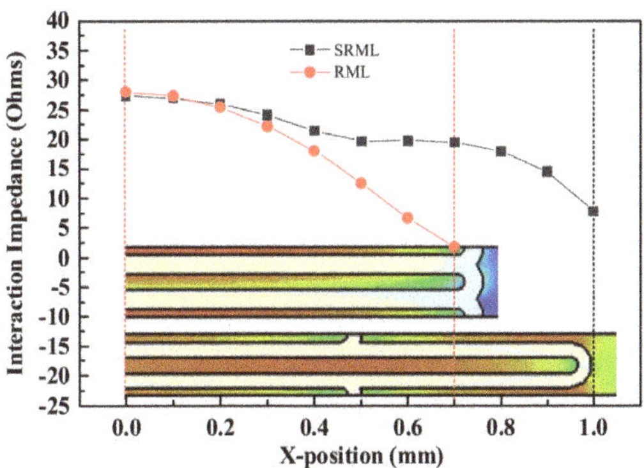

Figure 3. The variation of the interaction impedance with x position of the interaction impedance calculation line.

Figure 4 presents the curves of the normalized phase velocity and the interaction impedance with different values for the staggered distance of the rings. As the degree of interleaving of the rings increased, the normalized phase velocity and the coupling impedance decreased in the frequency band of 30 GHz to 40 GHz. Figure 5 shows the influence of the ring width on the normalized phase velocity and the coupling impedance of the SRML SWS. As the w_r increased, the normalized phase velocity and the bandwidth of the SRML SWS decreased. However, the interaction impedance slightly increased below the frequency of 40 GHz.

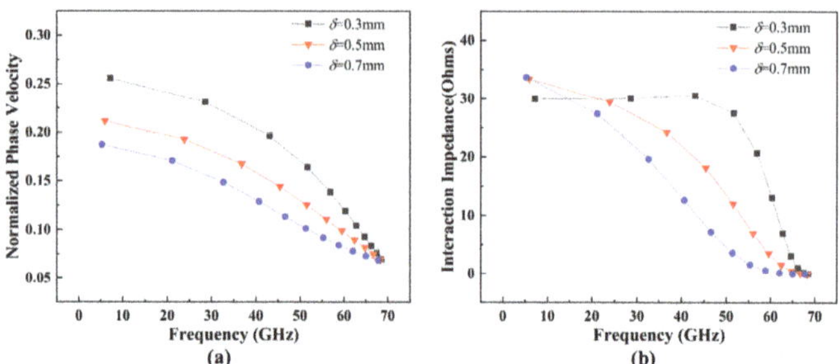

Figure 4. (**a**) Normalized phase velocity with different δ, (**b**) interaction impedance with different δ.

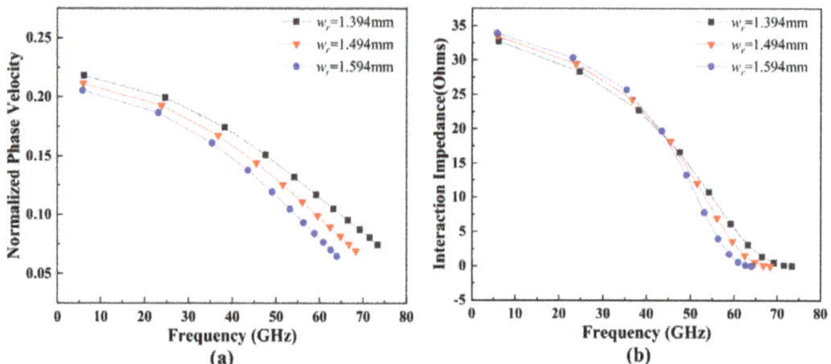

Figure 5. (**a**) Normalized phase velocity with different w_r, (**b**) interaction impedance with different w_r.

Figure 4 presents the curves of the normalized phase velocity and the interaction impedance with different values for the staggered distance of the rings. As the degree of interleaving of rings increased, the normalized phase velocity and the coupling impedance decreased in the frequency band of 30GHz to 40GHz. Figure 5 shows the influence of the ring width on the normalized phase velocity and the coupling impedance of the SRML SWS. As the w_r increased, the normalized phase velocity and the bandwidth of SRML SWS decreased. However, the interaction impedance slightly increased below the frequency of 40 GHz.

Through a series of simulation and optimizations, the value parameters of the SRML SWS were obtained, as shown in Table 1.

Table 1. The parameters of the SRML SWS.

Parameter	Value (mm)
p	0.128
w_r	1.494
w	0.04
s	0.034
δ	0.5
h_d	0.254
t	0.005
a	4
h_a	1.254

3. The Transmission Characteristics of the SRML SWS

The model of SRML SWS used for the transmission characteristics simulation was established in CST Microwave Studio, as shown in Figure 6. The model consists of 100 periods of SRML SWSs, 6 periods of transition section, microstrip probes, rectangular waveguides, and a metal shielding box. Through the microstrip probes, the input signal was converted from the TE_{10} mode in the rectangular waveguide to the quasi-TEM mode in the microstrip line. The material of the shielding box was copper. In the simulation, the conductivity of the copper was set as 2.8×10^7 S/m, the relative permittivity of the quartz glass was 3.75, and the loss tangent was 4×10^{-4}.

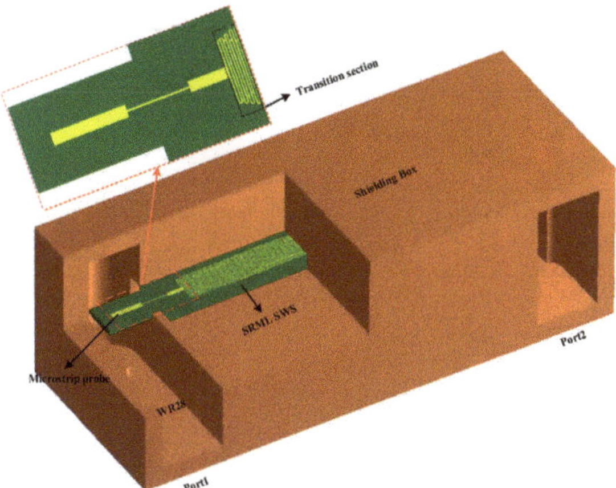

Figure 6. Schematic diagram of SRML SWS transmission model.

Figure 7 shows the reflection loss and the insertion loss of the SRML SWS with input/output waveguides. The reflection coefficient of the SRML SWS is less than −15 dB in the frequency range of 26 GHz to 40 GHz, and the insertion loss is around 12.5 dB at the center frequency of 32.5 GHz. The simulation results suggest that the designed SRML SWS has good transmission characteristics.

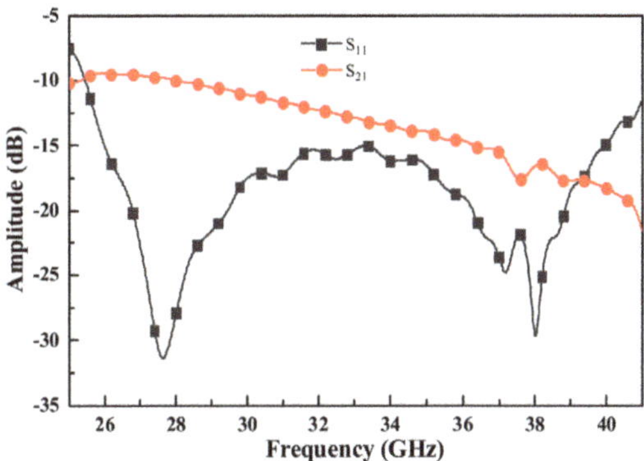

Figure 7. S-parameters of SRML SWS with input/output waveguide.

4. PIC Simulation of the SRML TWT

To reduce the simulation time, a model of the SRML SWS without input/output waveguide was established in CST Particle studio, as shown in Figure 8. The model consisted of 250 periods of the SRML SWS, whose length was around 32 mm. As analyzed in Section 2, since the adjacent rings of the SRML SWS were staggered placed, the metal line of the SRML SWS had a wider transverse width. So, a wider sheet electron beam could be used to feed DC energy for the beam–wave interaction process. In this simulation, the cross section of the sheet electron beam was 1.9 mm × 0.1 mm, and the beam current was 0.38 A, whose corresponding current density was 200 A/cm^2. The beam voltage was 9.7 kV, and the focusing magnetic field was a uniform magnetic field, whose amplitude was 0.8 Tesla.

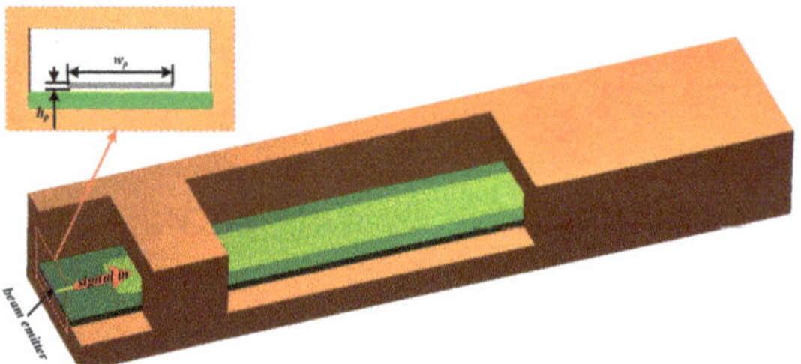

Figure 8. Schematic diagram of SRML TWT model for PIC simulation.

Figure 9 shows the output signal in the time domain at 32.5 GHz, and the input power is 1 W. After around 1.2 ns, the output power becomes stable with an amplitude of 25.4 V; the corresponding power and gain are 322.58 W and 25.09 dB, respectively. Additionally, from Figure 9, the output signal stayed stable in the simulation time of 10 ns, which means that no oscillation occurred.

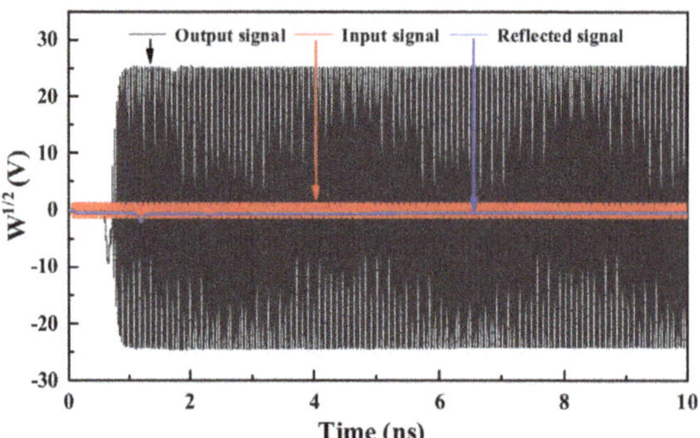

Figure 9. Output signals of the SRML TWT in time domain.

The electron beam trajectories and the phase space diagram after beam–wave interaction has been shown in Figure 10. It shows that the electron beam was well modulated, and there was an obvious transcendence phenomenon at the end of the TWT, indicating that the wave received plenty of energy from the electron beam. Figure 10 also shows that the electron beam smoothly passed through the entire SWS without hitting the dielectric substrate under the constraint of the magnetic field.

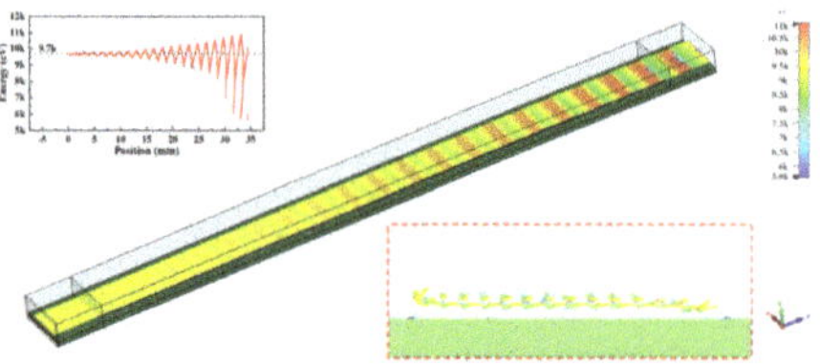

Figure 10. Electron beam bunching and phase space diagram.

Figure 11 shows the output power varied with frequency. The maximum output power was around 322.58 W with an electronic efficiency of 8.74% at 32.5 GHz. By slightly decreasing the working voltage, the SRML TWT could have an amplified output power of 26 GHz to 39 GHz; the output power was over 100 W in the frequency range of 27 GHz to 38 GHz and 200 W in the frequency range of 29 GHz to 37 GHz, which means the SRML TWT was a broadband TWT.

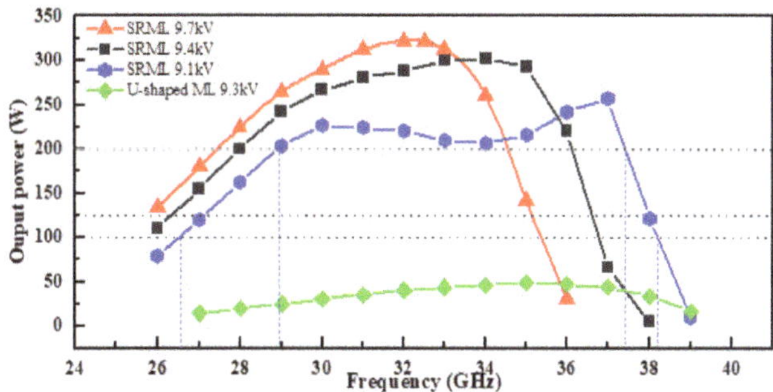

Figure 11. Spectrogram of output signal and reflected signal.

What is more, the output power of the U-shaped ML TWT, which had almost the same phase velocity as the SRML SWS at 32.5 GHz, was obtained, as shown by the green line in Figure 11. Although the 3 dB bandwidth of the SRML TWT was slightly narrower than that of the U-shaped ML TWT, the output power of the SRML TWT was far higher than that of the U-shaped ML TWT. The explanation for this is that the SRML SWS had a wider interaction area, which was almost 3.73 times that of the U-shaped ML SWS, so the SRML TWT could be driven by an electron beam with a higher beam current when the current density was the same.

5. Discussions

As a kind of structure that has been well studied in recent years, PCB (printed circuit board) [19], UV-LIGA [6,9], and DRIE [6] technologies have been used to obtain ML SWS successfully. This means SRML SWSs can be fabricated conveniently.

Taking UV-LIGA technology as an example, the processing of SRML SWSs is shown in Figure 12. The first step was to coat a layer of photoresist on the dielectric substrate, and then the cured photoresist was shielded with the specific mask and exposed under ultraviolet light, as shown in Figure 12b. Next, the exposed dielectric substrate was baked and developed to obtain a photoresist mold, as shown in Figure 12c. Finally, the dielectric substrate was metalized, and the required SRML SWS was obtained after removing the photoresist mold, as shown in Figure 12e.

For a TWT, the optical electron system was also very important. In this design, the electron beam adopted in the SRML TWT had a large aspect ratio, which was equal to 19. However, since the perveance and the beam current density were around 0.398 $uA/V^{3/2}$ and 200 A/cm^2, respectively, which are not very high, it could be realized through the common sheet beam electron gun. As the length of the slow-wave circuit was around 44 mm, to maintain the electron beam passing through the slow-wave circuit successfully, a uniform magnetic field will be designed; a uniform magnetic field with an amplitude of around 0.8 Tesla is not very hard to achieve. As shown in Figure 13, a uniform magnetic-focusing system was designed. The simulation result of the uniform magnetic is presented in Figure 14. The magnitude of the uniform magnetic field was around 0.81 Tesla, the length of the rising edge of the magnetic field was around 6.8 mm, and the length of the uniform magnetic field was more than 44.4 mm.

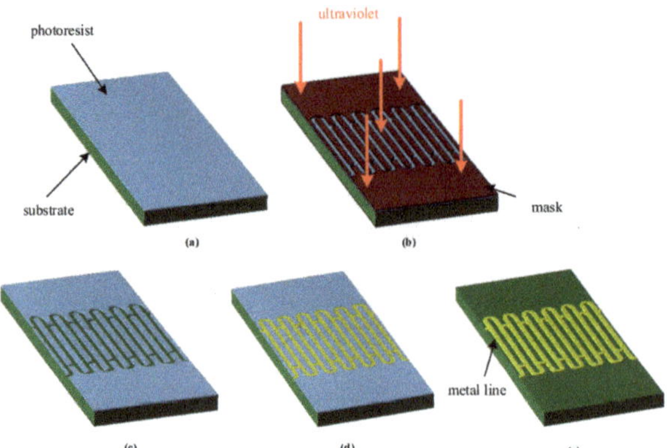

Figure 12. Diagram of processing SRML SWS by UV-LIGA technology. (**a**) covering photoresist on dielectric substrate, (**b**) lithography, (**c**) photoresist mold after developing, (**d**) metallization and (**e**) SRML SWS after removing photoresist.

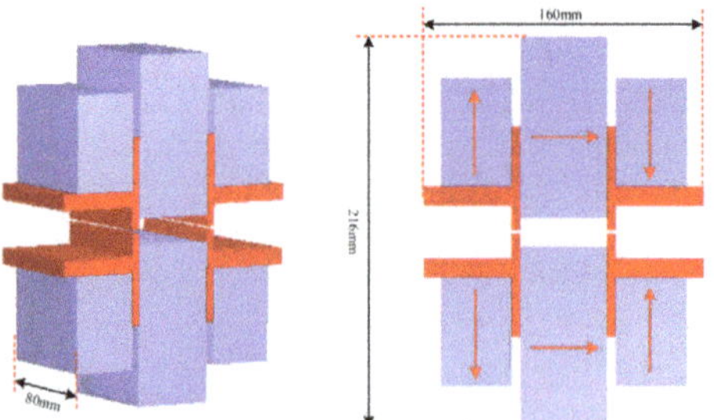

Figure 13. Schematic diagram of uniform magnetic field-focusing system and magnetization direction.

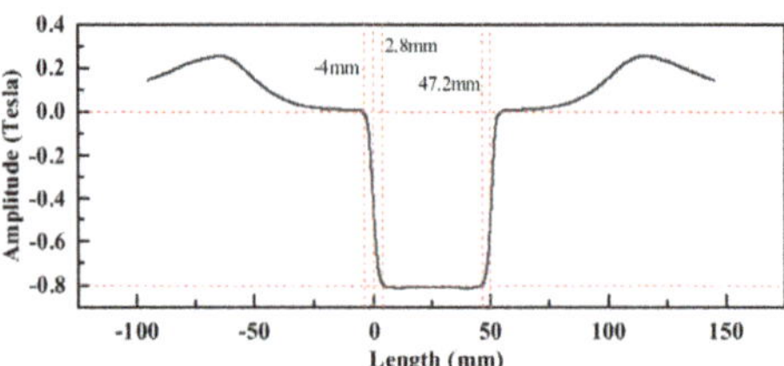

Figure 14. Variation of Z-component of the magnetic field.

6. Conclusions

To obtain a higher output power with a microstrip SWS, a SRML SWS has been designed and simulated based on RML SWS. By staggering the adjacent rings of a RML SWS, the transverse of the metal line was enlarged, which means that the SRML SWS had a larger area than the RML SWS for the beam–wave interaction. This feature caused the SRML TWT to have a higher electron beam current than that of the RML TWT when the current density remained the same. So, the output power will be higher. The simulation results of the interaction impedance show that the SRML SWS had a larger value in a wider area, which also indicates that a SRML can have a higher output power. The particle-in-cell simulation results show the designed SRML TWT had a maximum output power of 322.5 W at 38 GHz with an electronic efficiency of 8.74%, and the 3 dB bandwidth was over 7 GHz.

Author Contributions: Conceptualization, R.Y., J.X., and P.Y.; methodology, R.Y., G.G., and D.L.; validation, H.Y., H.W., and J.L.; writing—original draft preparation, J.Z. and D.J.; writing—review and editing, J.C., W.W., and L.Y.; supervision, Y.W.; project administration, G.Z., J.X., and L.Y. All authors have read and agreed to the published version of the manuscript.

Funding: This work was supported in part by the Basic Science Center Program of National Natural Science Foundation of China (NSFC) (Grant No. 61988102) and National Natural Science Foundation of China (NSFC) (Grant No. 61771117).

Institutional Review Board Statement: Not applicable.

Informed Consent Statement: Not applicable.

Data Availability Statement: The data presented in this study are available on request from the corresponding author.

Conflicts of Interest: The authors declare no conflict of interest.

References

1. Qiu, J.; Levush, B.; Pasour, J.; Katz, A.; Armstrong, C.; Whaley, D.; Tucek, J.; Kreischer, K.; Gallagher, D. Vacuum tube amplifiers. *IEEE Microw. Mag.* **2009**, *10*, 38–51. [CrossRef]
2. Ghosh, T.K.; Challis, A.J.; Jacob, A.; Bowler, D.; Carter, R.G. Improvements in Performance of Broadband Helix Traveling-Wave Tubes. *IEEE Trans. Electron Devices* **2008**, *55*, 668–673. [CrossRef]
3. Legarra, J.; Cusick, J.; Begum, R.; Kolda, P.; Cascone, M. A 500-W Coupled-Cavity TWT for Ka-Band Communication. *IEEE Trans. Electron Devices* **2005**, *52*, 665–668. [CrossRef]
4. Liu, S. Folded waveguide circuit for broadband MM wave TWTs. *Int. J. Infrared Millim. Waves* **1995**, *16*, 809–815. [CrossRef]
5. Zhang, D.G.; Yung, E.K.N.; Ding, H.Y. Dispersion Characteristics of a Novel Shielded Periodic Meander Line. *Microw. Opt. Technol. Lett.* **1996**, *12*, 1–5. [CrossRef]
6. Sengele, S.; Jiang, H.; Booske, J.H.; Kory, C.L.; Van Der Weide, D.W.; Ives, R.L. Microfabrication and Characterization of a Selectively Metallized W-Band Meander-Line TWT Circuit. *IEEE Trans. Electron Devices* **2009**, *56*, 730–737. [CrossRef]
7. Sumathy, M.; Augustin, D.; Datta, S.K.; Christie, L.; Kumar, L. Design and RF Characterization of W-band Meander-Line and Folded-Waveguide Slow-Wave Structures for TWTs. *IEEE Trans. Electron Devices* **2013**, *60*, 1769–1775. [CrossRef]
8. Socuellamos, J.M.; Dionisio, R.; Letizia, R.; Paoloni, C. Experimental Validation of Phase Velocity and Interaction Impedance of Meander-Line Slow-Wave Structures for Space Traveling-Wave Tubes. *IEEE Trans. Microw. Theory Tech.* **2021**, *69*, 2148–2154. [CrossRef]
9. Ryskin, N.M.; Rozhnev, A.G.; Starodubov, A.V.; Serdobintsev, A.A.; Pavlov, A.M.; Benedik, A.I.; Torgashov, R.; Torgashov, G.V.; Sinitsyn, N.I. Planar Microstrip Slow-Wave Structure for Low-Voltage V-Band Traveling-Wave Tube with a Sheet Electron Beam. *IEEE Electron Device Lett.* **2018**, *39*, 757–760. [CrossRef]
10. Chua, C.; Aditya, S. A 3-D U-Shaped Meander-Line Slow-Wave Structure for Traveling-Wave-Tube Applications. *IEEE Trans. Electron Devices* **2013**, *60*, 1251–1256. [CrossRef]
11. Bai, N.; Shen, M.; Sun, X. Investigation of Microstrip Meander-Line Traveling-Wave Tube Using EBG Ground Plane. *IEEE Trans. Electron Devices* **2015**, *62*, 1622–1627. [CrossRef]
12. Wang, S.; Gong, Y.; Wang, Z.; Wei, Y.; Duan, Z.; Feng, J. Study of the Symmetrical Microstrip Angular Log-Periodic Meander-Line Traveling-Wave Tube. *IEEE Trans. Plasma Sci.* **2016**, *44*, 1787–1793. [CrossRef]
13. Shen, F.; Wei, Y.; Yin, H.; Gong, Y.; Xu, X.; Wang, S.; Wang, W.; Feng, J. A Novel V-Shaped Microstrip Meander-Line Slow-Wave Structure for W-band MMPM. *IEEE Trans. Plasma Sci.* **2011**, *40*, 463–469. [CrossRef]
14. Shen, F.; Wei, Y.; Xu, X.; Liu, Y.; Huang, M.; Tang, T.; Duan, Z.; Gong, Y. Symmetric Double V-Shaped Microstrip Meander-Line Slow-Wave Structure for W-Band Traveling-Wave Tube. *IEEE Trans. Electron Devices* **2012**, *59*, 1551–1557. [CrossRef]
15. Liu, L.; Wei, Y.; Shen, F.; Zhao, G.; Yue, L.; Duan, Z.; Wang, W.; Gong, Y.; Li, L.; Feng, J. A Novel Winding Microstrip Meander-Line Slow-Wave Structure for V-Band TWT. *IEEE Electron Device Lett.* **2013**, *34*, 1325–1327. [CrossRef]
16. Chong, D.; Yanyu, W.; Qian, L.; Luqi, Z.; Guo, G.; Yubin, G. A dielectric-embedded microstrip meander line slow-wave structure for miniaturized traveling wave tube. *J. Electromagn. Waves Appl.* **2017**, *31*, 1938–1946. [CrossRef]
17. Ding, C.; Wei, Y.; Zhang, L.; Guo, G.; Wang, Y.; Zhang, M.; Lu, Z.; Gong, Y.; Wang, W.; Li, D.; et al. Beam-wave interaction study on a novel Ka-band ring-shaped microstrip meander-line slow wave structure. In Proceedings of the 2014 39th International Conference on Infrared, Millimeter, and Terahertz waves (IRMMW-THz), Tucson, AZ, USA, 14–19 September 2014; pp. 1–2. [CrossRef]
18. Ulisse, G.; Krozer, V. Investigation of a planar metamaterial slow wave structure for traveling wave tube applications. In Proceedings of the 2017 Eighteenth International Vacuum Electronics Conference (IVEC), London, UK, 24–26 April 2017; pp. 1–2. [CrossRef]
19. Guo, G.; Yan, Z.; Sun, Z.; Liu, J.; Yang, R.; Gong, Y.; Wei, Y. Broadband and Integratable 2 $\times$ 2 TWT Amplifier Unit for Millimeter Wave Phased Array Radar. *Electronics* **2021**, *10*, 2808. [CrossRef]

 *electronics* 

Article

Investigation on High-Efficiency Beam-Wave Interaction for Coaxial Multi-Beam Relativistic Klystron Amplifier

Limin Sun [1,2], Hua Huang [1], Shifeng Li [1,*], Zhengbang Liu [1], Hu He [1], Qifan Xiang [1,3], Ke He [1] and Xianghe Fang [1]

1 Science and Technology on High-Power Microwave Laboratory, Institute of Applied Electronics, China Academy of Engineering Physics, Mianyang 621900, China; slm14@163.com (L.S.); hhua0457@163.com (H.H.); liu9559@yeah.net (Z.L.); hyj0827@sina.com (H.H.); qfanxiang@163.com (Q.X.); hekegscaep@163.com (K.H.); fangxianghecn@163.com (X.F.)

2 Sichuan Institute of Industrial Technology, School of Electronic Information and Computer Engineering, Deyang 618500, China

3 School of Electronic Science and Engineering, University of Electronic Science and Technology of China (UESTC), Chengdu 610054, China

* Correspondence: lishifeng@alu.uestc.edu.cn

Abstract: To significantly improve the electronic efficiency of coaxial multi-beam relativistic klystron amplifier (CMB-RKA), the physical process of beam-wave interaction and parameters that affect efficiency was studied. First, the high efficiency of beam-wave interaction was discussed by simulating the efficiency versus the parameters (frequency of cavity, drift tube length between cavities, and external quality factor of output cavity), in the one-dimensional (1-D) large-signal simulation software. Moreover, the further physical process of beam-wave interaction was analyzed through simulating the current modulation factor and the number of particles at the entrance of the output cavity, in the three-dimensional (3-D) particle in cell simulation software. Last, with the optimal parameters in 3-D simulations, the CMB-RKA, which has 14 electron beams with a total current of 4.2 kA (14 $\times$ 300 A), can generate an output power of 1.02 GW with a saturation gain of 55.6 dB and an efficiency of 48.7%, when beam voltage is 500 kV, which indicated the CMB-RKA can achieve high efficiency for high-power microwave radiation.

Keywords: high-power microwave; relativistic klystron amplifier; high efficiency; multi-beam; beam-wave interaction

Citation: Sun, L.; Huang, H.; Li, S.; Liu, Z.; He, H.; Xiang, Q.; He, K.; Fang, X. Investigation on High-Efficiency Beam-Wave Interaction for Coaxial Multi-Beam Relativistic Klystron Amplifier. *Electronics* **2022**, *11*, 281. https://doi.org/10.3390/electronics11020281

Academic Editor: Yosef Pinhasi

Received: 9 December 2021
Accepted: 11 January 2022
Published: 17 January 2022

Publisher's Note: MDPI stays neutral with regard to jurisdictional claims in published maps and institutional affiliations.

1. Introduction

Relativistic klystron amplifier (RKA) is one of the promising high-power microwave (HPM) sources and is widely used in high-power radars, new accelerators, and new communication systems, because of its advantages in high-power, high-efficiency, stable-phase, and stable-amplitude of output power [1–3]. It has achieved an output power of 100 MW to several GW from the L-band to the Ka-band [4–10], and it is easy to obtain high radiation power in the low-frequency band, such as the 15-GW L-band RKA developed by Friedman et al. at the US Naval Laboratory [4].

After decades of development, RKA no longer pursues power improvement but focuses on miniaturization, high efficiency and low power consumption [11]. The high efficiency of RKA can reduce the volume and consumption of pulsed power sources, which is conducive to the miniaturization, stability, and reliability of the HPM system. Therefore, theoretical and experimental research on the physical mechanism of high-efficiency beam-wave interaction is regarded as a high priority in the future. At the same time, High-Efficiency International Klystron Activity (HEIKA) was established internationally driven by the next generation of large particle accelerators, such as Compact Linear Collider [12], Future Circular Collider [13], and International Linear Collider. Under the promotion of HEIKA, a series of methods to improve efficiency was proposed, such as Core Oscillation

Method (COM) [14], Bunching-Alignment-Collecting (BAC) [15], Cluster Center Stabilization Method (CSM) [16], Two-Stage Structure [17], etc. The essence of these methods is to achieve a core of the electron beam whose electron phases are the same, which leads to improvement in the efficiency of beam-wave interaction. Baikov A. Y. predicted that the RF electronic efficiency can be 90% by the COM method in theory [14], and the efficiency in MAGIC 2D simulation increased from the original 82.7% up to 84.6% by the COM method in high-power klystron amplifier [18]. However, for a relativistic klystron amplifier driven by intense electron beams, its electronic efficiency is limited to about 40%, because of strong space-charge forces [19,20].

In this paper, we investigate the physical mechanism of high-efficiency beam-wave interaction based on a five cavities coaxial multi-beam relativistic klystron amplifier (CMB-RKA), in which (i) the perveance is reduced by the multi-beam to improve efficiency; (ii) the fundamental-mode coaxial cavity enhances the characteristic impedance (R/Q) and coupling coefficient (M) to achieve high-efficiency beam-wave interaction. Moreover, the parameters affecting the beam-wave interaction, which are high-frequency parameters, structural parameters, and operating parameters of CMB-RKA, are studied through one-dimensional (1-D) and three-dimensional (3-D) simulations. They are adopted to the design of the CMB-RKA. As a result, the 1.02 GW output power is achieved, corresponding with an efficiency of 48.7% and a saturation gain of 55.6 dB. It is believed that the presented results would be of great interest to the design and development of high-efficiency RKA.

2. Physical Design

The structure of CMB-RKA is shown in Figure 1, and its operating principle can be described as follows. An RF input signal injects into the input cavity to excite an operating mode. The intense relativistic electron beams are slightly velocity modulated across the gap of the input cavity. Then, the beams are density-modulated in the drift tubes to present bunching core and peripheral electrons. In addition, the velocity is modulated again by the idler cavities, which benefits to push peripheral electrons into the bunch. Last, as the beams travel down the drift tubes and the idler cavities to the output cavity, the electrons are decelerated with converting their kinetic power into RF power. The efficiency depends on the modulated current exciting, which is closely related to the electron beams and cavities. Thus, the physical design is developed first, including the parameters of electron beam and cavities.

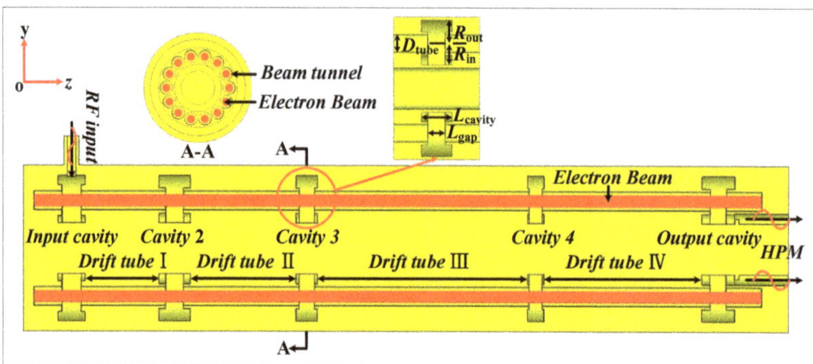

Figure 1. y-z sectional view of the S-band CMB-RKA.

2.1. Design of Electron Beam Parameters

In klystron, the electronic efficiency is defined as [21]

$$\eta = \frac{1}{2}\frac{P_n}{P_0} = \frac{1}{2}\frac{V_n}{V_0}\frac{I_n}{I_0} \tag{1}$$

In Equation (1), P_0, V_0, I_0 are the power, voltage, current of the electron beam, respectively; P_n, V_n, I_n are the nth harmonic power, voltage, current of the modulated electron beam, respectively. V_n and I_n can be expanded by a cosine function, and the factor $1/2$ comes from the time average of the square of the cosine function in an RF cycle. Usually, the current modulation factor is defined as $m_n = I_n/I_0$. When m_n is 2, the efficiency is 100%, which means the density of modulated beam is well represented as a δ function. However, other factors should be considered, such as space-charge force, to counteract the beam bunching, which leads that the efficiency cannot reach a theoretical 100%. Generally, the space-charge force is characterized directly by perveance ($P = I_0 V_0^{-3/2}$) [14], and Ref. [22] indicates that the efficiency is growing with a reduction of perveance. Thus, to obtain high efficiency, perveance should be small for a single electron beam.

Additionally, the multi-beam klystron uses several metal wall-separated channels to prevent communication between electron beams so that the operating voltage can be reduced significantly [23]. This means low perveance and miniaturized devices. Consequently, the structure of the multi-beam is the best option for high efficiency. The efficiency was calculated by perveance in MBK, as shown in the equation ($\eta = 0.78 - 0.16P$) [24], where the unit of perveance P is μK, and the maximum efficiency can be 78%. Here, we propose a CMB-RKA with 14 beamlets of 500 kV and 300 A. The perveance of a beamlet is 0.85 μK and the total perveance of the multi-beam electron beam is 11.9 μK, where the theoretical efficiency of the CMB-RKA is 64.4%.

2.2. Design of Cavity Parameter

Multiple cavities are adopted to achieve high gain and high efficiency. In addition, Ref [14] researched the maximal efficiency of two multi-beam klystrons (7-beams and 42-beams) that their efficiency all increased non-linearly as the number of cavities increased. When the number of cavities is greater than 5, the efficiency increases slowly. Therefore, we set the number of cavities to be 5. The fundamental mode of the coaxial resonant cavity is adopted in the CMB-RKA, as shown in Figure 1. The effects of structural parameters (L_{cavity}, L_{gap}, R_{out}, and R_{in}) of the cavity on high-frequency parameters (M, f, and R/Q) are simulated in 3-D EM analysis software(CST Studio Suite) [25], and the results are shown in Figure 2. We can adjust the structural parameters to obtain the appropriate high-frequency parameters, which can be used in 1-D and 3-D simulations.

In addition, the equivalent circuit of the coaxial resonant cavity is an *RLC* circuit, as shown in Figure 3a. L and C represent the high-frequency characteristics of the toroidal and parallel-plate portion in the cavity, R represents the electronic beam load, external load, and cavity wall loss. When the operating frequency is smaller than the frequency of the cavity, the circuit is inductive, therefore the phase of V will lag behind the phase of I_i [26]. Figure 3b shows the situation where the phase difference ($\Delta\varphi = \varphi(V) - \varphi(I_i)$) between the gap voltage of cavity (V) and the electron current (I_i) is 90°, whose electron beams is under a sinusoidal perturbation. In this case, fast-velocity electrons are decelerated, and slow-velocity electrons are accelerated by gap voltage (V) so that the peripheral electrons can be pushed into the bunching core. Thus, the detuning cavity operating in fundamental mode is applied to bunch peripheral electrons and obtain the high efficiency of the device.

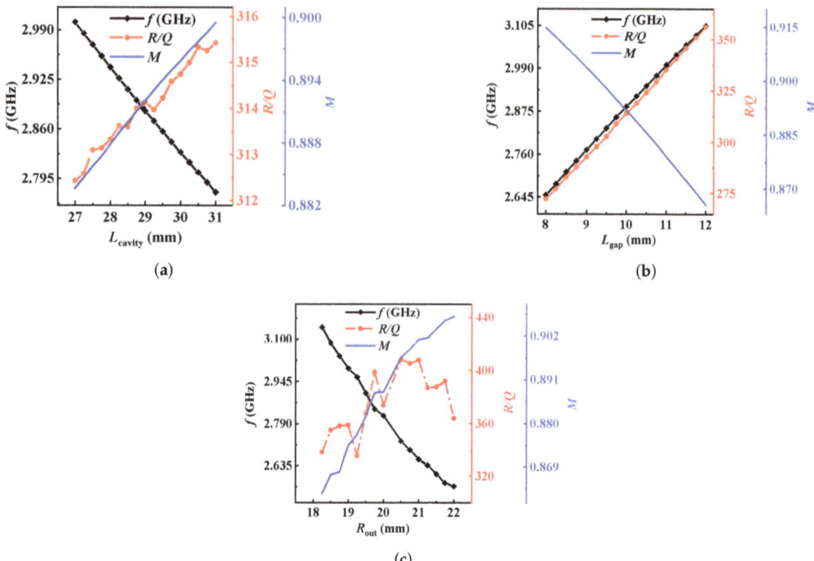

Figure 2. Frequency (f), coupling coefficient (M), and characteristic impedance (R/Q) of the cavity as functions of (**a**) axial length of the cavity (L_{cavity}), (**b**) gap length of the cavity (L_{gap}), and (**c**) Rout when R_{in}/R_{out} is a constant.

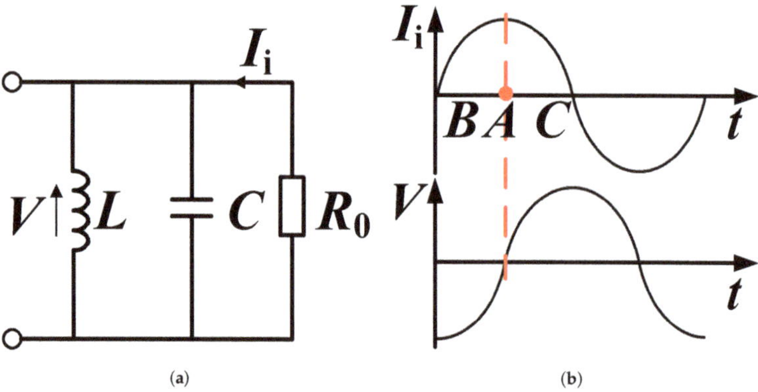

Figure 3. (**a**) Equivalent circuit of the fundamental-mode coaxial cavity. (**b**) Electron current (I_i) and gap voltage of cavity (V) under a sinusoidal perturbation (Electrons in point A is in the electron bunching core, Electrons in point B is fast-velocity electrons, and Electrons in point C is slow-velocity electrons.).

3. Beam-Wave Interaction

The efficiency of a klystron is determined by a set of parameters that can be divided into two groups. The parameters in the first group are beam voltage, beam current, number of beams, injection power, and operating frequency. The parameters in the second group are R/Q, M, and frequency of the cavity (f_n), external quality factors of the input and output cavities (Q_{ext}), number of the cavity and drift tube length between the cavities (l_n). In this section, the high-efficiency of beam-wave interaction was discussed by some parameters of these, in 1-D large-signal numerical simulation software (AJDISK) and 3-D particle in cell (PIC) simulation software (CHIPIC).

3.1. 1-D Large-Signal Research on the Physical Process of Beam-Wave Interaction

AJDISK is a 1-D large-signal numerical simulation software developed at SLAC and based on the disc model, which is adapted to simulate klystrons with a single cylindrical and sheet beam [27]. Thus, when it was used to analyze the high-efficiency beam-wave interaction of CMB-RKA, the equivalent must be done. In AJDISK simulation, the R/Q is 14 times the CMB-RKA and the beam current is 300 A that 1/14 times the CMB-RKA, the other parameters, such as M, beam voltage, f_n, beam radius, drift tube radius, input power, Q_{ext}, and l_n, keep the same with the CMB-RKA. Moreover, the single variable method was taken in the whole process of the research. It means that only one parameter was changed whereas the others were fixed.

3.1.1. Effect of Cavity Frequency on Beam-Wave Interaction

According to Section 2.2, the detuning cavities can help the peripheral electrons move into the bunching core. That is because the phase difference between the gap voltage of the cavity and the modulation electron current increases with the frequency of the cavity. As a result, a weak velocity modulation of electrons within the bunching core can be, but the antibunching electrons. And the bunching core will oscillate for the space-charge force and the weak velocity modulation so that the antibunching electrons will be kicked into the bunching core due to a stronger velocity modulation. Thus, for high efficiency, the initial bunching core must be generated first, by the cavities that their frequency are near the operating frequency, and the collecting with the peripheral electrons is later by the detuning cavities. As shown in Figure 4, f_1 and f_2 are near the operation frequency, f_3 and f_4 obviously increases, especially, f_4 increase to hundreds of MHz comparing the operating frequency. It indicates that the input cavity and cavity 2 play the main role on producing the initial bunching core, and the peripheral electrons are gathered at the bunching core by cavity 3 and 4 dominantly. Moreover, the frequency of the output cavity acting a decelerated structure is near the operating frequency.

Furthermore, the effective range of cavity frequency is different. As shown in Figure 4, the frequency range of the cavity 2 is minimum, and the frequency range of the cavity 4 is maximum. Therefore, the frequency of the cavity 2 cannot deviate too much in design and experiment. On the other hand, the frequency of the input cavity should be stabilized at the design frequency of 2.88 GHz because the electron beams entering the input cavity are uniform.

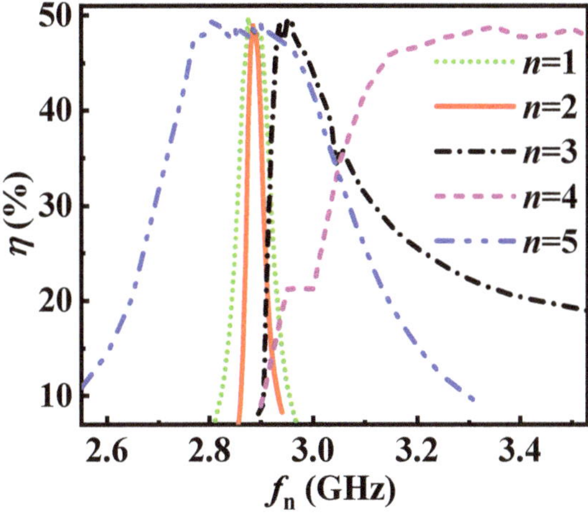

Figure 4. Efficiency versus frequency of the cavity f_n (f_1, f_2, f_3, f_4, f_5 are frequency of the input cavity, cavity 2, cavity 3, cavity 4, and output cavity, respectively.).

3.1.2. Effect of the Drift Tube Length between Cavities on Beam-Wave Interaction

According to Section 3.1.1, the input cavity and cavity 2 play the main role in producing the initial bunching core, thus drift tube length I (l_1) and II (l_2) are relatively short (Figure 5). Furthermore, as the generation of the initial bunching core, the density of charge within the bunching core increases, and the space-charge force will start to push the electrons away from the bunching core. Results that the bunch electrons will travel non-monotonic toward the bunching core, and the antibunching electrons without the debunching space-charge force will travel monotonic toward the bunch. This is beneficial to high efficiency, but it requires a substantial increase in the drift tube length of the later stage. As shown in Figure 5, drift tube length III (l_3) and IV (l_4) is longer. In particular, when drift tube length III (l_3) is 550 mm, the efficiency can be close to 60%.

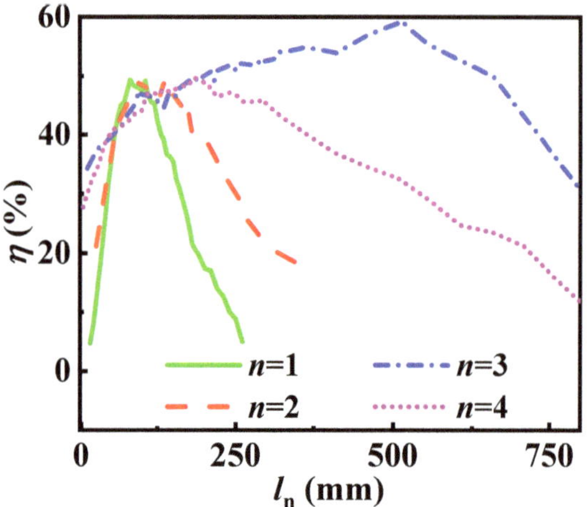

Figure 5. Efficiency versus drift tube length l_n (l_1, l_2, l_3 and l_4, are the drift tube length I, II, III, IV, respectively.).

3.1.3. Effect of the Q_{ext} of the Output Cavity on Beam-Wave Interaction

In the output cavity, the Q_{ext} was used to describe its coupling with the external coupler waveguide, and relate closely to its gap voltage to decelerate the modulated electrons sufficiently. Generally, the gap voltage of the output cavity is roughly equivalent to the beam voltage when the electron beam is well-modulated. If the gap voltage is too high the partial beams will turn around, resulting in a decline in the efficiency. Therefore, it is necessary to find an appropriate Q_{ext} of the output cavity to obtain a high efficiency. As shown in Figure 6, the best Q_{ext} of the output cavity is 10, and its range is very limited.

3.2. 3-D Pic Research on the Physical Process of Beam-Wave Interaction

In Section 3.1, we obtain the key parameters that affect the efficiency; however, the further physical process is not present from those parameters. The 3-D PIC simulation is more in line with the actual model of CMB-RKA and the software we used is CHIPIC which was developed at the University of Electronic Science and Technology of China and was proved a validity electromagnetic PIC code [28]. It can clearly analyze the bunching process of the electron beam and the physical nature of the high-efficiency beam-wave interaction. In this section, we investigate the effect of the frequency of cavity 2 (f_2) and the drift tube length III (l_3) on the beam bunching, by observing and analyzing the nth ($n = 1, 2, 3$) harmonic current (I_1, I_2, I_3) of modulated beam and the number of particles at the entrance of the output cavity.

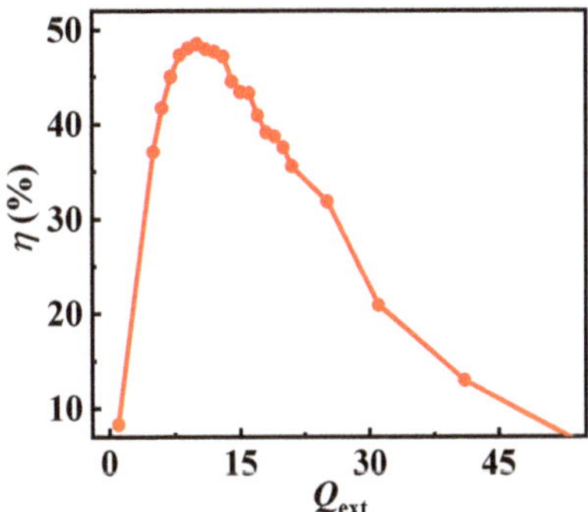

Figure 6. Efficiency versus Q_{ext} of the output cavity.

3.2.1. Effect of the Frequency of the Cavity 2 on Beam-Wave Interaction

From Section 3.1, it is known that the frequency of the cavity 2 (f_2) should be near the operating frequency, and its range is very limited. In this section, we simulated the modulation current (I_n, n = 1, 2, 3) and the number of particles (N) at the entrance of the output cavity in 3-D PIC. It should be noted that N is physical particles within each macroparticle in CHIPIC. Additionally, the analysis of I_{nmax}/I_0 (n = 1, 2, 3) and N is illustrated in Figure 7. In Figure 7a, the first harmonic maximum current modulation factor (I_{1max}/I_0) is over 1.6 as f_2 increases from 2.885 to 2.893 GHz, and the changes in I_{nmax}/I_0 (n = 2, 3) coincide well with I_{1max}/I_0. This suggests that the beam-wave interaction efficiency is high within this range of f_2. In Figure 7b, N is largest when f_2 is 2.889 GHz, proving that the peripheral electrons kicked into the bunching core is the most. Consequently, for the CMB-RKA, the frequency of the cavity 2 is 2.889 GHz and I_{1max}/I_0 is 1.78 respectively.

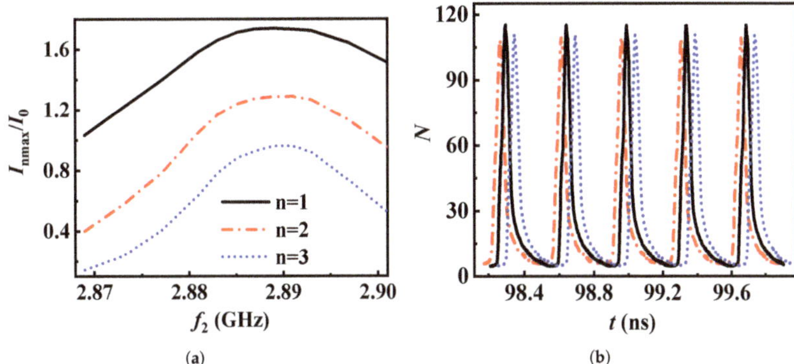

Figure 7. (**a**) Maximum current modulation factor versus frequency of the cavity 2 (I_{nmax}/I_0 is the maximum modulation current of the nth harmonic, n = 1, 2, 3.). (**b**) Number of particles at the entrance of the output cavity (f_2 of short dash dot line is 2.885 GHz, f_2 of solid line is 2.889 GHz, f_2 of short dot line f_2 is 2.893 GHz.).

3.2.2. Effect of the Drift Tube Length III on Beam-Wave Interaction

After the velocity modulation in the first of three cavities and the density modulation in the drift tube, the bunching core has been formed. To improve beam-wave interaction efficiency, the drift tube length III (l_3) should be long enough to collect the peripheral electrons. When l_3 is from 180 to 240 mm, I_{1max}/I_0 is over 1.75, and the changes in I_{nmax}/I_0 (n = 2, 3) coincide well with I_{1max}/I_0 (Figure 8a). In addition, N is largest when l_3 is 210 mm (Figure 8b). Therefore, the drift tube length III (l_3) is 210 mm and I_{1max}/I_0 is 1.81, respectively. Figure 8c shows the distribution of the modulation current along the z-axis. When Z = 591 mm, I_n (n = 1, 2, 3) is the largest. Hence, we can set the output cavity at Z = 591 mm to obtain the highest RF power.

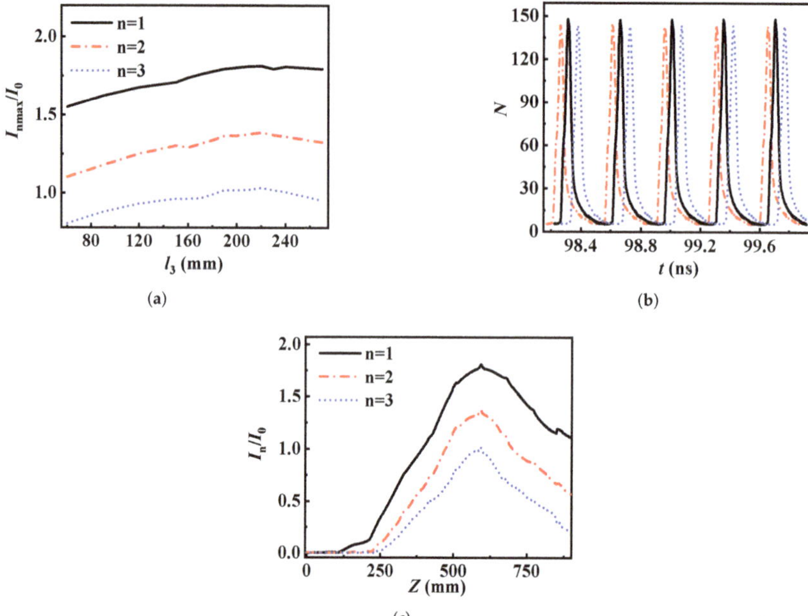

Figure 8. (**a**) Maximum current modulation factor versus drift tube length III (I_{nmax} is the maximum modulation current of the nth harmonic, n = 1, 2, 3). (**b**) Number of particles at the entrance of the output cavity (l_3 of short dash dot line is 180 mm, l_3 of solid line is 210 mm, l_3 of short dot line is 240 mm). (**c**) Current modulation factor along the length of the CMB-RKA.

4. Application in S-Band CMB-RKA

After 1-D and 3-D calculations, we select some high-efficiency parameters which are shown in Table 1, and the corresponding efficiency in 1-D is 52%. According to the parameters, the 3-D model of CMB-RKA was established in the CHIPIC (Figure 1), and the operation mode of its cavity is TM_{01}. In the simulation, we validate the application of the above investigations and design a high-efficiency CMB-RKA for HPM radiation at S-band.

Table 1. Parameters of 1-D calculation.

Parameters	Input Cavity	Cavity 2	Cavity 3	Cavity 4	Output Cavity
R/Q	265	240	199	201	228
M	0.8556	0.8686	0.8826	0.851	0.8753
Q_{ext}	25	95,000	95,000	95,000	9
f_n (MHz)	2880	2889	2944	3287	2865
Z (mm)	0	95	215	425	591

The Q_{ext} of the output cavity is closely related to the extraction of output microwave energy, and there is an optimal value (Figure 9). As Q_{ext} of the output cavity increases, the number of electrons that run back towards the negative z-axis increases, leading to a decrease in output power. Consequently, the Q_{ext} of the output cavity is set as 8.

Due to the difference in the algorithms adopted by AJDISK and CHIPIC, the saturation gain of the device will be different. It is necessary to simulate the saturation gain of S-band

CMB-RKA at different input power through CHIPIC. As shown in Figure 10, the output power becomes saturated when the input power is 2.8 kW. If the input power increases, electrons turn around seriously in the output cavity, causing a decrease in the output power (Figure 10). Thus, the input power is taken as 2.8 kW, corresponding saturation gain of 55.6 dB, and the efficiency of 48.7%. Comparing the efficiency between 1-D and 3-D, the difference is only 3%, which proves that the results of the 1-D large-signal simulation are credible.

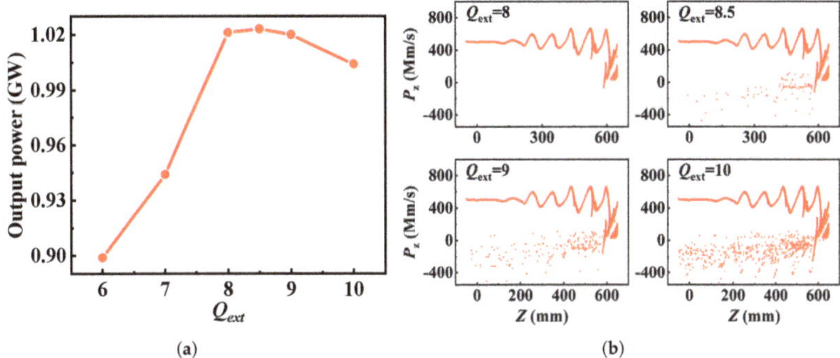

Figure 9. (**a**) Output power versus the Q_{ext} of the output cavity. (**b**) Axial momentum of the electrons along the length of the CMB−RKA.

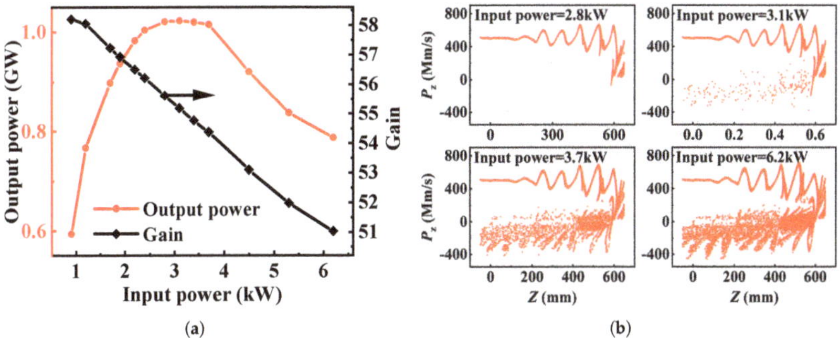

Figure 10. (**a**) Output power and gain versus input power. (**b**) Axial momentum of the electrons along the length of the CMB−RKA.

After optimization, the waveform of output power is shown in Figure 11. The output power and efficiency are 1.02 GW and 48.7%, respectively, indicating that the S-band CMB-RKA can operate stably with high efficiency.

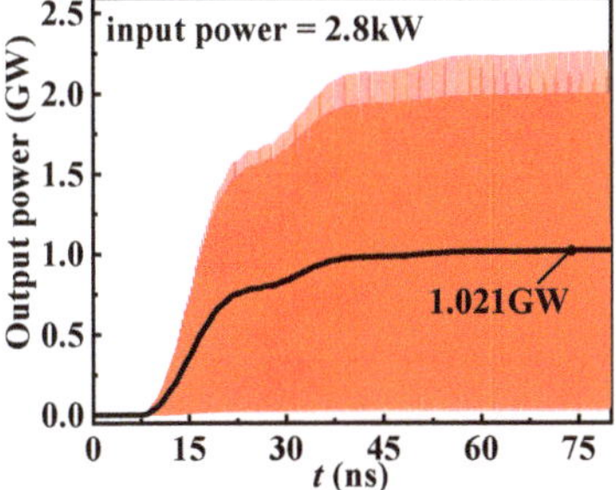

Figure 11. Output power waveform.

5. Conclusions

For high electronic efficiency, the 1-D software AJDISK and the 3-D software CHIPIC are adopted to study the beam-wave interaction mechanism of the S-band CMB-RKA. It is found that the essence of improving efficiency is to allow as many peripheral electrons as possible to enter the bunching core so that the phase of all the electrons is the same. In other words, when the m_n is 2, the theoretical efficiency is up to 100%. In the physical design, we adopt multi-beam, multi-cavity, and detuning cavity for high efficiency. In the study of high-efficiency beam-wave interaction, we systematically calculated the electronic efficiency versus the parameters which are frequency of the cavity, drift tube length between the cavities, and external quality factor of the output cavity in 1-D. Furtherly, we simulated the m_n (n = 1, 2, 3) and the number of particles at the entrance of output cavity in 3-D. From these simulations and studies, some conclusions are as follows: a. To achieve high efficiency, the frequency of the idler cavity should gradually increase from the first one to the last one, and the frequency range of the cavity 2 should be very small. b. A longer drift tube length can be beneficial to gain higher efficiency. c. The external quality factor of the output cavity should be the best. Finally, when the input power is 2.8 kW, the saturation gain and efficiency of S-band CMB-RKA are 55.6 dB and 48.7% respectively with 1.02 GW output power, and the bunching length is 591 mm.

Author Contributions: Conceptualization, L.S. and S.L.; methodology, L.S., S.L. and H.H. (Hu He); validation, S.L., H.H. (Hua Huang) and Z.L.; writing—original draft preparation, L.S.; writing—review and editing, L.S., S.L., Q.X., K.H. and X.F.; supervision, H.H. (Hua Huang) and Z.L.; funding acquisition, S.L. All authors have read and agreed to the published version of the manuscript.

Funding: This research was funded by the Science and Technology on High-Power Microwave Laboratory Fund under Grant No.6142605190201 and JCKYS2021212013.

Institutional Review Board Statement: Not applicable.

Informed Consent Statement: Not applicable.

Data Availability Statement: The data presented in this study are available on request from the corresponding author.

Conflicts of Interest: The authors declare no conflict of interest.

References

1. Friedman, M.; Pasour, J.; Smithe, D. Modulating electron beams for an X band relativistic klystron amplifier. *Appl. Phys. Lett.* **1997**, *71*, 3724–3726. [CrossRef]
2. Gold, S.H.; Nusinovich, G.S. Review of high-power microwave source research. *Rev. Sci. Instruments* **1998**, *68*, 3945. [CrossRef]
3. Benford, J.; Swegle, J.A.; Schamiloglu, E. Chapter 9. Klystrons and Reltons. In *High Power Microwaves*; CRC Press: Boca Raton, FL, USA, 2007; pp. 375–379.
4. Friedman, M.; Krall, J.; Lau, Y.; Serlin, V. Efficient generation of multigigawatt rf power by a klystronlike amplifier. *Rev. Sci. Instruments* **1990**, *61*, 171–181. [CrossRef]
5. Huang, H.; Chen, Z.; Li, S.; He, H.; Yuan, H.; Liu, Z.; Lei, L. Investigation on pulse-shortening of S-band, long pulse, four-cavity, high power relativistic klystron amplifier. *Phys. Plasmas* **2019**, *26*, 033107. [CrossRef]
6. Wu, Y.; De-Kui, Z.; Yong-Dong, C. Design of a C-band relativistic extended interaction klystron with coaxial output cavity. *Chin. Phys. C* **2015**, *39*, 077005. [CrossRef]
7. Liu, Z.; Huang, H.; Jin, X.; Li, S.; Wang, T.; Fang, X. Investigation of an X-Band Long Pulse High-Power High-Gain Coaxial Multibeam Relativistic Klystron Amplifier. *IEEE Trans. Electron Devices* **2019**, *66*, 722–728. [CrossRef]
8. Yang, F.; Dang, F.; He, J.; Zhang, X.; Ju, J. A Large Signal Theory of Multiple Cascaded Bunching Cavities for High-Efficiency Triaxial Klystron Amplifier. *Electronics* **2021**, *10*, 1284. [CrossRef]
9. Dang, F.; Ju, J.; Yang, F.; Ge, X.; Zhang, J.; He, J.; Zhang, X. Design and preliminary experiment of a disk-beam relativistic klystron amplifier for Ku-band long-pulse high power microwave radiation. *Phys. Plasmas* **2020**, *27*, 113101. [CrossRef]
10. Li, S.; Duan, Z.; Huang, H.; Basu, B.N.; Wang, F.; Liu, Z.; He, H.; Wang, X.; Wang, Z.; Gong, Y. Input and Output Couplers for an Oversized Coaxial Relativistic Klystron Amplifier at Ka-Band. *IEEE Trans. Electron Devices* **2019**, *66*, 2758–2763. [CrossRef]
11. Huang, H.; Wu, Y.; Liu, Z.-B.; Yuan, H.; He, H.; Li, L.L.; Li, Z.-H.; Jin, X.; Ma, H.-G. Review on high power microwave device with locked frequency and phase. *Acta Phys. Sin.* **2018**, *67*, 88402. [CrossRef]

12. Aicheler, M.; Burrows, P.; Draper, M.; Garvey, T.; Lebrun, P.; Peach, K.; Phinney, N.; Schmickler, H.; Schulte, D.; Toge, N. *A Multi-TeV Linear Collider Based on CLIC Technology: CLIC Conceptual Design Report*; Report; CERN: Geneva, Switzerland, 2014. [CrossRef]
13. Koratzinos, M. FCC-ee accelerator parameters, performance and limitations. *Nucl. Part. Phys. Proc.* **2016**, *273–275*, 2326–2328. [CrossRef]
14. Baikov, A.Y.; Marrelli, C.; Syratchev, I. Toward High-Power Klystrons with RF Power Conversion Efficiency on the Order of 90 *IEEE Trans. Electron Devices* **2015**, *62*, 3406–3412. [CrossRef]
15. Egorov, R.V.; Guzilov, I.A.; Maslennikov, O.Y.; Savvin, V.L. BAC-Klystrons: A New Generation of Klystrons in Vacuum Electronics. *Mosc. Univ. Phys. Bull.* **2019**, *74*, 38–42. [CrossRef]
16. Hill, V.C.R.; Marrelli, C.; Constable, D.; Lingwood, C. Particle-in-cell simulation of the third harmonic cavity F-Tube klystron. In Proceedings of the 2016 IEEE International Vacuum Electronics Conference (IVEC), Monterey, CA, USA, 19–21 April 2016; pp. 1–2. [CrossRef]
17. Teryaev, V.E.; Shchelkunov, S.V.; Hirshfield, J.L. 90% Efficient Two-Stage Multibeam Klystron: Modeling and Design Study. *IEEE Trans. Electron Devices* **2020**, *67*, 5777–5782. [CrossRef]
18. Constable, D.A.; Lingwood, C.; Burt, G.; Baikov, A.Y.; Syratchev, I.; Kowalcyzk, R. MAGIC2-D simulations of high efficiency klystrons using the core oscillation method. In Proceedings of the 2017 Eighteenth International Vacuum Electronics Conference (IVEC), London, UK, 24–26 April 2017; pp. 1–2. [CrossRef]
19. Li, S.; Huang, H.; Duan, Z.; Basu, B.N.; Liu, Z.; He, H.; Wang, Z. Demonstration of a Ka Band Oversized Coaxial Multi Beam Relativistic Klystron Amplifier for High Power Millimeter Wave Radiation. *IEEE Electron Device Lett.* **2021**, *43*, 131–134. [CrossRef]
20. Liu, Z.; Huang, H.; Jin, X.; Lei, L.; Zhu, L.; Li, L.; Li, S.; Yan, W.; He, H. Investigation of the phase stability of an X-band long pulse multibeam relativistic klystron amplifier. *Phys. Plasmas* **2016**, *23*, 093110. [CrossRef]
21. Lemke, R.W.; Clark, M.C.; Marder, B.M. Theoretical and experimental investigation of a method for increasing the output power of a microwave tube based on the split-cavity oscillator. *J. Appl. Phys.* **1994**, *75*, 5423–5432. [CrossRef]
22. Liu, Z.; Zha, H.; Shi, J.; Chen, H. Study on the Efficiency of Klystrons. *IEEE Trans. Plasma Sci.* **2020**, *48*, 2089–2096. [CrossRef]
23. Gelvich, E.; Borisov, L.; Zhary, Y.; Zakurdayev, A.; Pobedonostsev, A.; Poognin, V. The new generation of high-power multiple-beam klystrons. *IEEE Trans. Microw. Theory Tech.* **1993**, *41*, 15–19. [CrossRef]
24. Beunas, A.; Faillon, G. A high power long pulse high efficiency multi beam klystron. In Proceedings of the 5th MDK Workshop, CERN, Geneva, Switerland, 26 April 2001.
25. CST Computer Simulation Technology GmBH, Germany. 2016, CST Mircrowave Studio. Available online: http://www.cst.com (accessed on 10 January 2022).
26. Gilmour, A. *Klystrons, Traveling Wave Tubes, Magnetrons, Crossed-Field Amplifiers, and Gyrotrons*; Artech House Publishers: Norwood, MA, USA, 2011.
27. Jensen, A.; Fazio, M.; Neilson, J.; Scheitrum, G. Developing Sheet Beam Klystron Simulation Capability in AJDISK. *IEEE Trans. Electron Devices* **2014**, *61*, 1666–1671. [CrossRef]
28. Zhou, J.; Liu, D.; Liao, C.; Li, Z. CHIPIC: An Efficient Code for Electromagnetic PIC Modeling and Simulation. *IEEE Trans. Plasma Sci.* **2009**, *37*, 2002–2011. [CrossRef]

 *electronics* 

Article

Design and Experiments of the Sheet Electron Beam Transport with Periodic Cusped Magnetic Focusing for Terahertz Traveling-Wave Tubes

Changqing Zhang [1], Pan Pan [1], Xueliang Chen [1], Siming Su [1], Bowen Song [1], Ying Li [1], Suye Lü [2], Jun Cai [1], Yubin Gong [3] and Jinjun Feng [1,*]

[1] National Key Laboratory of Science and Technology on Vacuum Electrics, Beijing Vacuum Electrics Research Institute, Beijing 100015, China; c.q.zhang@163.com (C.Z.); p-pan@hotmail.com (P.P.); chenxueliang0611@126.com (X.C.); 18645093816@163.com (S.S.); songbob6968@sina.com (B.S.); ly18046510463@sina.com (Y.L.); caijun@sdu.edu.cn (J.C.)

[2] Institute of Nano-Photoelectronics and High Energy Physics, Beijing Institute of PetroChemical Technology, Beijing 102617, China; lvsuye@bipt.edu.cn

[3] National Key Laboratory of Science and Technology on Vacuum Electronics, University of Electronic Science and Technology of China, Chengdu 610054, China; ybgong@uestc.edu.cn

* Correspondence: fengjinjun@tsinghua.org.cn

Abstract: The successful transport of a sheet electron beam under the periodic cusped magnet (PCM) focusing at the terahertz frequencies is reported. The sheet beam with a current density of 285 A/cm^2 is intended for the developing G-band sheet-beam traveling-wave tube (TWT) whose operating voltage is nominally 24.5 kV. A beamstick was developed to validate the design of the electron optics system, which is considered as the most challenging part for developing a sheet-beam device. A beam transmission ratio of 81% is achieved over a distance of 37.5 mm at a cathode voltage of −25.0 kV. The total current and the collector current were measured to be 125 and 102 mA, respectively. The experimental results are promising, demonstrating that the PCM scheme is capable of focusing a high-current-density sheet beam and hence can find use in the terahertz TWTs, offering the advantages of compact size and light weight.

Keywords: Terahertz TWTs; sheet electron beam; PCM focusing; transmission of the sheet beam; vacuum electronic devices

Citation: Zhang, C.; Pan, P.; Chen, X.; Su, S.; Song, B.; Li, Y.; Lü, S.; Cai, J.; Gong, Y.; Feng, J. Design and Experiments of the Sheet Electron Beam Transport with Periodic Cusped Magnetic Focusing for Terahertz Traveling-Wave Tubes. *Electronics* **2021**, *10*, 3051. https://doi.org/10.3390/electronics10243051

Academic Editor: Yahya M. Meziani

Received: 9 November 2021
Accepted: 3 December 2021
Published: 7 December 2021

1. Introduction

The sheet electron beam has been long considered for enhancing the performance of vacuum electronic devices (VEDs). However, the development of a sheet-beam device has been hindered for decades due to the difficulties in transport known as the "diocotron instability" [1,2]. Significant breakthroughs have been made in the last ten years. Several prototype devices have been successfully developed [3–7], which demonstrates the great potential of the sheet-beam technology, especially in power enhancement. Those inventions in many ways represent state-of-the-art technologies of developing VEDs except that most of them are based on solenoidal magnetic fields generated by the permanent magnets [8,9] whose volume and weight increase dramatically as the magnetic field peak and the transport distance increased. This is a major concern for using the sheet beam in a practical scenario such as the airborne radars and communication applications. Therefore, the periodic permanent magnet (PPM) focusing of the sheet beam is still attractive, although its application is found to be limited to transporting a sheet beam with low voltage and high current density [9].

The PCM focused sheet-beam TWT was first demonstrated at Ka band with beam transmission of 93% with a voltage of 24.3 kV [10]. At the same frequency band, a PCM-focused high-power sheet-beam TWT was recently reported [11] with a CW power output of up to 3 kW, along with a transmission of 97% at a voltage of 30.9 kV and a current of 0.82 A. In addition, a W-band PCM-focused sheet-beam extended-interaction oscillator (EIO) was demonstrated with a transport of 94.4% at a high operating voltage of ~47 kV [12]. However, it remains a challenge for the PCM scheme to transport a low-voltage high-current-density sheet beam for the terahertz TWTs. The reasons come from both the insufficient capacity of the magnet material itself and the structural limitations from the reduced size of the circuit. Thus, compromises have to be made between the desired the high current and the achievable magnetic fields. We recently performed a theoretical review of the PCM focusing, along with the design of a magnetic field system [13]. As a key step of developing the sheet-beam TWT, a beamstick was developed. This paper presents a complete design of the electron optical system. The experimental results are also reported for the first time.

2. Design of the Electron Gun

The design of the electron gun was based on the CST (Computer Simulation Technology) Particle Tracking Solver. The nominal design parameters are listed in Table 1. The simulation model is shown in Figure 1. Basically, this is a Pierce-type gun with the exception that the cross sections of the cathode and the focusing electrode are conformal ellipses, as shown in Figure 1b. To facilitate the assembly, the shape of the electrodes is designed as simply as possible. Consequently, both the cathode emission surface and the anode entrance are planar. In the tracking simulation, we enable the gun iteration option to ensure the convergence of the emitting current. The effects of the space-charge field and the self-magnetic field of the beam are also considered in the simulation. The 2D particle monitors were deployed along the beam transport direction with an interval of 0.1 mm. As a result, the position and other statistical information of the particles can be derived by these monitors.

Table 1. Design parameters of the sheet-beam electron gun.

Parameters	Value (Unit)
Cathode voltage V_0	24.5 kV
Emission current I_0	0.135 A
Cathode size	0.85×0.65 mm^2
Beam waist size [1]	0.6×0.11 mm^2
Tunnel size	0.8×0.2 mm^2

[1] The cross section of the beam is an ellipse. Thus, the dimension of beam is determined by the major axis and the minor axis of the ellipse.

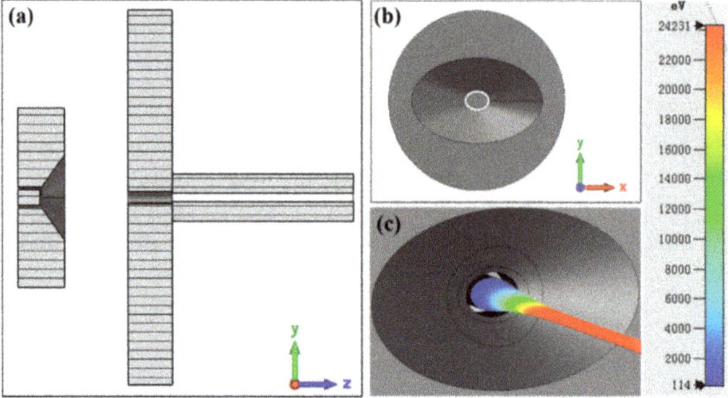

Figure 1. The CST model of the sheet-beam electron gun: (**a**) the cross section of the model, (**b**) the front view of the cathode and focus electrode, and (**c**) the beam formation with the elliptical focus electrode.

Figure 2 shows the convergence of the emitting current. The simulation predicted a current of 135 mA at a cathode voltage of 24.5 kV. The resulting cathode loading is 31.1 A/cm^2, which exceeds the capacity of the ordinary M-type cathodes. Thus, the scandate cathode is used.

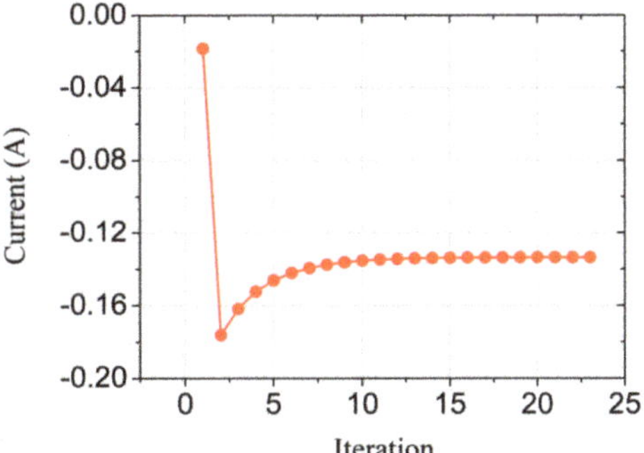

Figure 2. The convergence of the emitting current.

Figure 3 shows the electrostatic trajectory envelopes of the sheet beam. We can see that the gun provides an excellent laminar property for the sheet beam. The electron beam is compressed in both directions. However, the major contribution to the compression ratio comes from the vertical direction (Y). The position of the beam waist in the vertical direction is basically the same as that in the horizontal direction, which is about 7.8 mm. The compression ratios of the horizontal and vertical directions are 1.37 and 6.5, respectively, corresponding to an overall area compression ratio of 8.9. Figure 4 demonstrates the phase space of the compressing process where the two cross sections correspond to the emission surface and the beam waist, respectively. We can see that the sheet beam maintains its shape well during the compressing process, and no over-compressed phenomenon is observed at the edges of the beam.

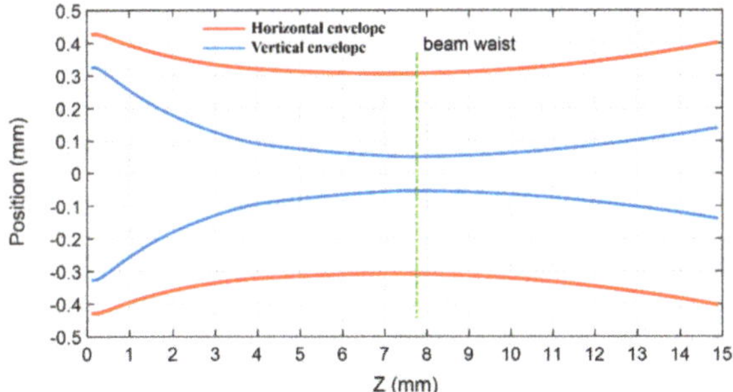

Figure 3. The envelopes of the electrostatic trajectories of the sheet beam.

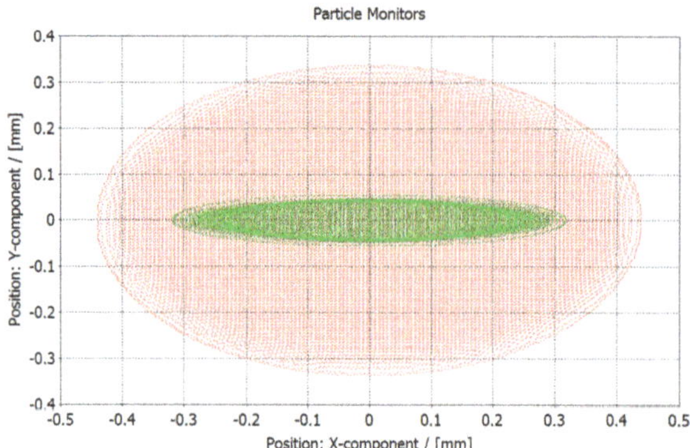

Figure 4. Compression of the sheet beam from emission surface to the waist.

3. Design of the PCM Focusing System

The stable transmission of the sheet beam under the periodically focusing magnetic fields requires the two-plane focusing, i.e., both the vertical and the horizontal directions need to be considered. Several focusing schemes can realize that. Here, a modified offset-pole-piece PCM scheme [13] is used. The model of the focusing system is shown in Figure 5. The arrows in the magnets denote the direction of the magnetization. The CST Magnetostatic Solver was used for the design of the magnet system. The parameters of the magnetic system are listed in Table 2. In the simulation, the iron material is assumed for the pole pieces with a nonlinear *B–H* relationship.

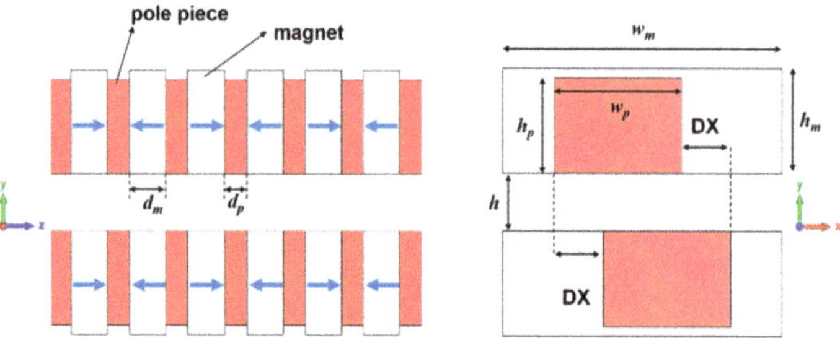

Figure 5. Schematic diagram of the PCM focusing system.

Table 2. Parameters of the magnet system.

Parameters	Value (Unit)
B_r	1.09 T
L_m	6.4 mm
h	3.0 mm
$w_m \times h_m \times d_m$	15 × 5.5 × 2.0 mm^3
$w_p \times h_p \times d_p$	6.8 × 5.0 × 1.2 mm^3

Figure 6a shows the distribution of the on-axis magnetic field B_z. It can be seen that a peak of 0.38 T is achieved. Figure 6b plots the distribution of B_y along the longitudinal direction at the beam sides X = ±0.3 mm and Y = 0. This magnetic field is denoted as the side magnetic field $B_{y,\text{side}}$. We can see that it is essentially a *dc* field, which is different from B_z. Thus, the mechanism for horizontal focusing is different from that of the vertical focusing. The theory predicted that a value of 10^{-3} T for B_y is required to counter the space-charge defocusing effect from $E_{sp,x}$. We can see from Figure 6b that the design meets the requirement.

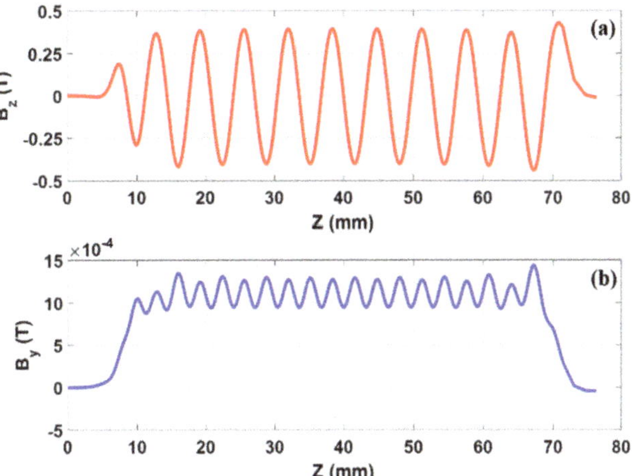

Figure 6. Design results of the focusing magnetic fields: (**a**) the on-axis distribution of B_z, (**b**) the side magnetic field $B_{y,\text{side}}$ at X = ±0.3 mm and Y = 0.

The unique distribution of $B_{y,\text{side}}$ is highly sensitive to the position, as is illustrated in Figure 7, where dY denotes the position deviation from the Y = 0 plane with X = 0.3 mm constant. It can be seen that the distribution of $B_{y,\text{side}}$ varies dramatically. For this reason, a means of tuning the side magnetic field is critical in practice.

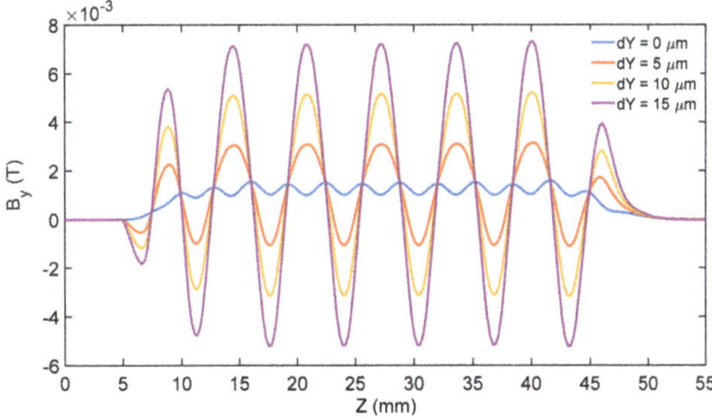

Figure 7. Sensitivity analysis of the side magnetic field $B_{y,\text{side}}$.

By importing the designed magnetic fields into the electron gun model, a complete electron optical simulation can be achieved. Figure 8 shows the formation and transmission of the sheet beam under the designed focusing magnetic fields. It can be seen that the sheet beam is well confined in both directions. Moreover, although the horizontal envelope of the sheet beam is highly sensitive to the side magnetic field, the vertical focusing is nearly independent of the side focusing. This is the basis of tuning the side magnetic field in practice.

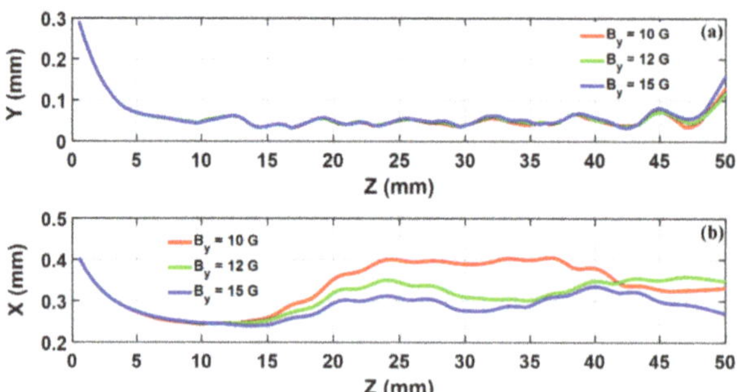

Figure 8. Simulations of the sheet beam transmission under the designed focusing magnetic fields with the side magnetic field $B_{y,\mathrm{side}}$ varied: (**a**) the trajectories envelopes in vertical direction and (**b**) the trajectories envelopes in horizontal direction.

In light of the importance of the magnetic field system, efforts have been made to measure the magnetic field distributions. The measuring platform is demonstrated in Figure 9a. This is a three-dimensional auto-measuring platform, which can provide a position-shifted accuracy of 10 μm and a magnetic field measurement accuracy of 10^{-6} T. The results are shown in Figure 9b,c with simulation results compared. Note that since the space between the two magnet arrays is too small to insert the probe in terms of the design value of h = 3.0 mm, we have to enlarge the space. As a result, the peak of the magnetic fields was reduced. However, we can see that the measured distribution of B_z is in good agreement with the simulation. Although discrepancy can be observed in B_y, the agreement is fair considering the sensitivity of the side magnetic field.

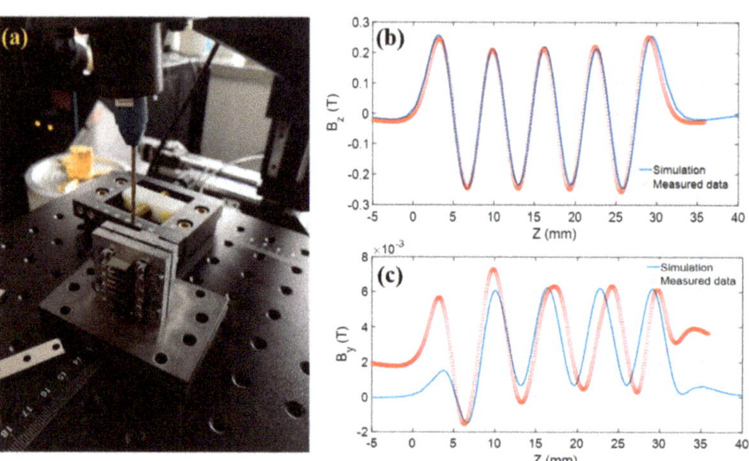

Figure 9. (**a**) The magnetic field measurement platform, (**b**) the measured axial magnetic field distribution B_z, and (**c**) the measured side magnetic field B_y.

4. Experimental Results

Experiments have been carried out to verify the design of the sheet-beam electron optical system. A beamstick was assembled without the need for RF input/output coupler, where the interaction circuit is replaced by a rectangular tunnel of 0.75 × 0.2 mm^2. The tunnel size is consistent with that of an actual interaction circuit. The transmission distance is 37.5 mm from the anode entrance to the collector. Such a distance is sufficient to support an amplification up to 18 dB.

Figure 10 demonstrates the beamstick in the testing site. The performance of the beamstick was measured under the pulse mode with pulse durations of 10–50 μs and repetition rates of 100 Hz. The body and the collector currents are measured by the ammeters and simultaneously confirmed with the oscilloscope. Figure 11 shows current waveforms for different pulse durations.

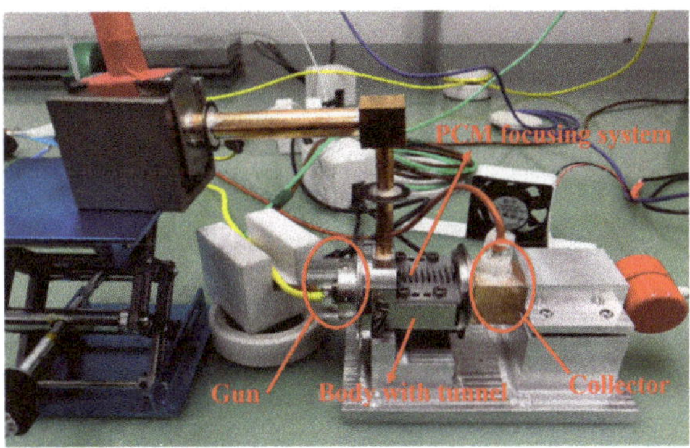

Figure 10. The beamstick in the testing site.

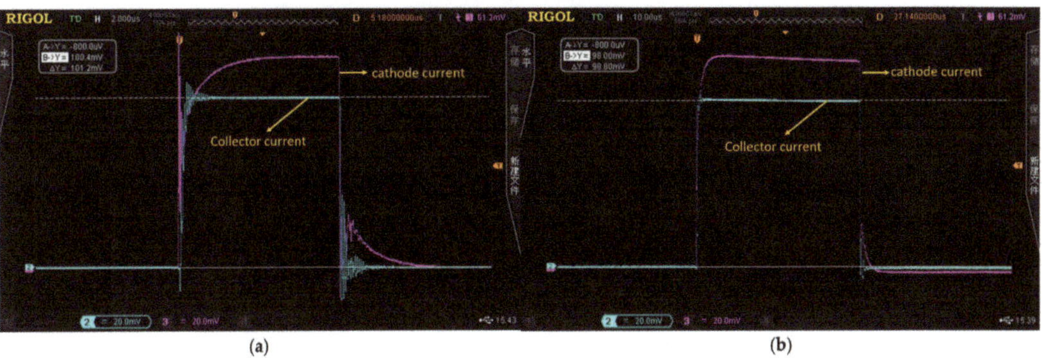

(**a**) (**b**)

Figure 11. The current waveforms measured by oscilloscope with (**a**) pulse durations of 10 μs and (**b**) pulse durations of 50 μs. Both cases have the same repetition rates of 100 Hz.

Figure 12 plots the measured cathode emission current. At 1050 °C, corresponding to a filament heating power ~4.4 W, the emission current was measured to be 125 mA, corresponding to a cathode loading of 28.8 A/cm^2. The current is slightly lower than the design value of 135 mA due to the biased voltage applied to the focus electrode. By reducing the biased voltage, the emission can approach 146 mA. Thus, the scandate cathode can provides sufficient emission for the sheet-beam gun.

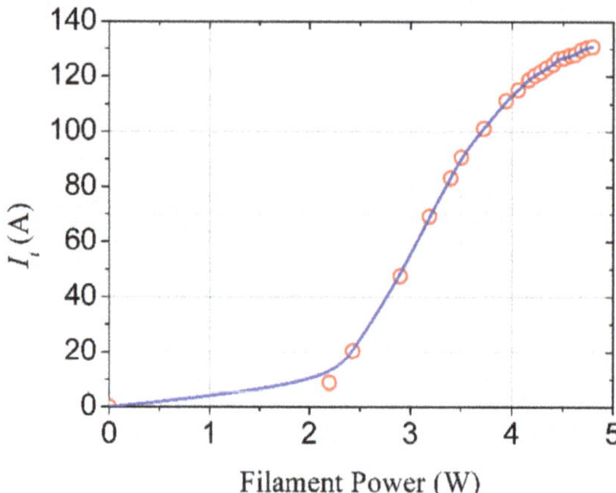

Figure 12. The emission current of the Scandate cathode as a function of the filament power.

Figure 13 shows the total current I_t and the collector current I_c as a function of the cathode voltage. It can be seen that the total current I_t increases almost linearly with the increase of the cathode voltage. The collector current I_c is small and rises slowly when the voltage is less than 18 kV, but it increases fast after that. An optimal transport efficiency over 81% was obtained at a voltage of −25 kV, which is slightly higher than the design value of −24.5 kV. The corresponding total current and collector current are 125 and 102 mA, respectively.

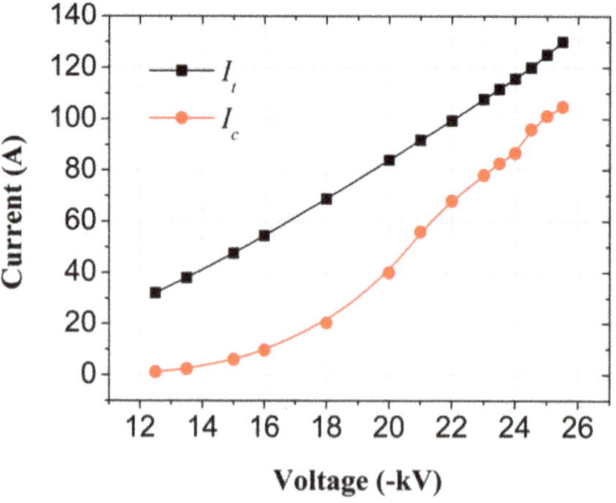

Figure 13. The measured currents as a function of the cathode voltage.

From the measured currents, we can estimate that the cutoff voltage of the PCM system is about 19 kV. Such a high cutoff voltage, which is close to the nominal operating voltage (24.5 kV), indicates that the PCM system may approach its limit of capacity. That is why the voltage corresponding to the optimal transport efficiency is higher than the nominal operating voltage. Due to the difficulty in achieving a higher peak of the magnetic field system, the voltage has to be increased to achieve the optimal transport efficiency.

Figure 14 shows the effect of the biased focus voltage (BFV) on the currents and the transport efficiency. It can be seen that both the collector current and the transport efficiency are significantly increased as the BFV increased before it reaches −85 V. As the BFV goes further, the increase of the transport efficiency is mainly attributed to the decrease of the total current. Due to the thermal expansion effect of the cathode and the inevitable errors in assembling, the required BFV is found to be much higher than expected. This is also the reason that the total current is lower than the design value at the nominal voltage of 24.5 kV.

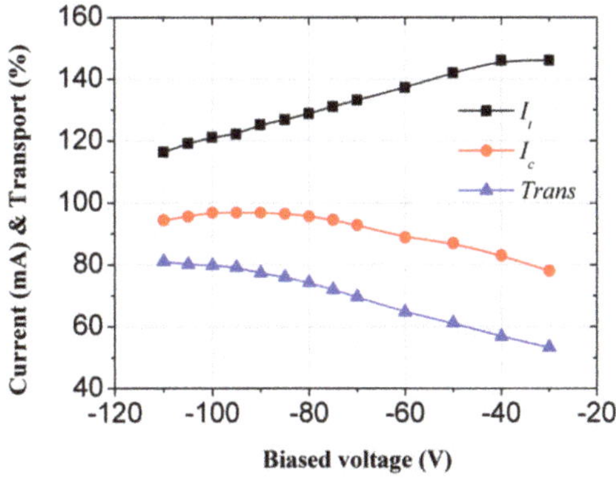

Figure 14. The measured current and the transport efficiency as a function of the biased focus voltage.

Table 3 compared the beam parameters and the transport efficiencies of the sheet-beam TWTs, which were reported with the experimental results. To make the comparison meaningful for different frequencies, the normalized transport distance was used, which is defined as L/λ_0, where $\lambda_0 = c/f_0$, L is the actual transmission distance, f_0 is the central frequency, and c is the speed of light in vacuum. We specify f_0 as 14, 35, 44, 94, and 220 GHz for the Ku, Ka, Q, W, and G band. It can be seen clearly from the table that this paper demonstrated the highest level in the cathode emitting capacity, beam current density, and normalized transport distance among the PCM focused TWTs. In particular, the transmission current density is nearly five times higher than that of the Ka-band TWT. Considering the more stringent requirements for tolerance in terahertz frequencies, we can conclude that a substantial breakthrough has been made in the long-distance transmission of the PCM focused sheet beam. The focusing with a permanent magnet solenoid allows a higher current density because a very large magnetic field (0.85~1 T) can be achieved. However, the disadvantages in volume and weight limit the use of a permanent magnet solenoid in TWTs.

Table 3. Comparison of the electron optical parameters of the sheet-beam TWTs, which were reported with experimental results.

Devices	Band	Focusing Scheme	Voltage (kV)	Current (A)	Beam Size and Cathode Size	Cathode Loading (A/cm^2)	Current Density (A/cm^2)	Normalized Transport Distance	Transport Efficiency (%)
This paper	G	PCM	24.5	0.135	0.60×0.11 mm^2 0.85×0.65 mm^2	31.1	285	27.5	81
TWT [10]	Ka	PCM	24.3	0.8	3.2×0.6 mm^2 3.2×1.8 mm^2	17.68	53.05	13.16	93
TWT [11]	Ka	PCM	30.9	0.82	4.0×0.4 mm^2 [1] 10:1	6.366	63.66	17	97
TWT [14]	Ku	PCM	34	4	8.0×0.9 mm^2 10:1	7.073	70.73	14	81
TWT [14]	Q	PCM	30	1.2	3.0×0.4 mm^2 10:1	12.73	127.3	17	92
Beamstick [3,5,8]	W/Ka	Solenoid [2]	20	4	4×0.25 mm^2 4×10 mm^2	10	400	6.27	98.5
TWT [4]	G	Solenoid	22.88	0.23	0.7×0.1 mm^2 0.86×1.06 mm^2	38	438	29.35	78.3

[1] The dimensions of the cathode cannot be found but the compression ratios were given. [2] Permanent magnet solenoid.

5. Conclusions

Efforts have been made in the design, fabrication, assembly, and measurement to achieve a PCM-focused sheet electron beam for the terahertz TWTs. A beamstick was developed and tested. A transport efficiency over 81% was measured with a total current of 125 mA and a collector current of 102 mA. The total current is slightly lower than the design value due to an increased focus voltage. Although such a current is still at an intermediate level for a sheet beam, it is two times higher than that of a circular beam with PPM focusing at the same frequencies. Thus, the performance enhancement of the TWT with a sheet beam can be expected.

Author Contributions: Conceptualization, C.Z. and J.F.; methodology, C.Z., P.P., and J.C.; software, S.L.; validation, C.Z., P.P., and J.C.; data curation, X.C., S.S., B.S., and Y.L.; writing—original draft preparation, C.Z.; writing—review and editing, J.F. and C.Z.; supervision, J.F.; project administration P.P., J.C and Y.G. All authors have read and agreed to the published version of the manuscript.

Funding: This research was funded by the Natural Science Foundation of China (Grant No. 62131006, 61921002, 61831001, 92163204) and the Beijing Natural Science Foundation (Grant No. 1192007), The Scientific Research Project of Beijing Educational Committee under Grant KM202010017010.

Conflicts of Interest: The authors declare no conflict of interest.

References

1. Cutler, C.C. Instability in Hollow and Strip Electron Beams. *J. Appl. Phys.* **1956**, *27*, 1028–1029. [CrossRef]
2. Booske, J.H.; McVey, B.D.; Antonsen, T.M. Stability and confinement of nonrelativistic sheet electron beams with periodic cusped magnetic focusing. *J. Appl. Phys.* **1993**, *73*, 4140–4155. [CrossRef]
3. Pasour, J.; Wright, E.; Nguyen, K.T.; Balkcum, A.; Wood, F.N.; Myers, R.E.; Levush, B. Demonstration of a Multikilowatt, Solenoidally Focused Sheet Beam Amplifier at 94 GHz. *IEEE Trans. Electron. Devices* **2014**, *61*, 1630–1636. [CrossRef]
4. Gamzina, D.; Himes, L.G.; Barchfeld, R.; Zheng, Y.; Popovic, B.K.; Paoloni, C.; Choi, E.; Luhmann, N.C. Nano-CNC Machining of Sub-THz Vacuum Electron Devices. *IEEE Trans. Electron. Devices* **2016**, *63*, 4067–4073. [CrossRef]
5. Pershing, D.E.; Myers, R.E.; Levush, B.; Nguyen, K.T.; Abe, D.K.; Wright, E.; Larsen, P.B.; Pasour, J.; Cooke, S.; Balkcum, A.; et al. Demonstration of a Wideband 10-kW Ka-Band Sheet Beam TWT Amplifier. *IEEE Trans. Electron. Devices* **2014**, *61*, 1637–1642. [CrossRef]
6. Cusick, M.; Atkinson, J.; Balkcum, A.; Caryotakis, G.; Gajaria, D.; Grant, T.; Meyer, C.; Lind, K.; Perrin, M.; Scheitrum, G.; et al. X-Band Sheet Beam Klystron (XSBK). In Proceedings of the 2009 IEEE International Vacuum Electronics Conference, Rome, Italy, 28–30 April 2009; pp. 296–297. [CrossRef]
7. Zhao, D.; Lu, X.; Liang, Y.; Yang, X.; Ruan, C.; Ding, Y. Researches on an X-Band Sheet Beam Klystron. *IEEE Trans. Electron. Devices* **2013**, *61*, 151–158. [CrossRef]
8. Pasour, J.; Nguyen, K.; Wright, E.; Balkcum, A.; Atkinson, J.; Cusick, M.; Levush, B. Demonstration of a 100-kW Solenoidally Focused Sheet Electron Beam for Millimeter-Wave Amplifiers. *IEEE Trans. Electron. Devices* **2011**, *58*, 1792–1797. [CrossRef]
9. Nguyen, K.T.; Pasour, J.A.; Antonsen, T.M.; Larsen, P.B.; Petillo, J.J.; Levush, B. Intense Sheet Electron Beam Transport in a Uniform Solenoidal Magnetic Field. *IEEE Trans. Electron. Devices* **2009**, *56*, 744–752. [CrossRef]
10. Shi, X.; Wang, Z.; Tang, T.; Gong, H.; Wei, Y.; Duan, Z.; Tang, X.; Wang, Y.; Feng, J.; Gong, Y. Theoretical and Experimental Research on a Novel Small Tunable PCM System in Staggered Double Vane TWT. *IEEE Trans. Electron. Devices* **2015**, *62*, 4258–4264. [CrossRef]
11. Wang, J.; Wan, Y.; Li, X.; Liu, Q.; Li, H.; Yao, Y.; Zheng, Q.; Wu, Z.; Jiang, W.; Liu, G.; et al. Continuous Wave Operation of a Ka-Band Broadband High-Power Sheet Beam Traveling-Wave Tube. *IEEE Electron. Device Lett.* **2021**, *42*, 1069–1072. [CrossRef]
12. Wang, J.; Li, X.; Rui, L.; Liu, Z.; Liu, G.; Jiang, W.; Wu, Z.; Hu, Y.; Luo, Y. Experimental Study of a 6 kW W-band PCM Focused Sheet Beam EIO. In Proceedings of the 2019 International Vacuum Electronics Conference (IVEC), Busan, Korea, 28 April–1 May 2019; pp. 1–2.
13. Zhang, C.; Feng, J.; Cai, J.; Pan, P.; Su, S.; Gong, Y. Focusing of the Sheet Electron Beam with Two-Plane Periodic Cusped Magnetic System for Terahertz TWTs. *IEEE Trans. Electron. Devices* **2021**, *68*, 3056–3062. [CrossRef]
14. Wang, J.; Liu, G.; Shu, G.; Zheng, Y.; Yao, Y.; Luo, Y. The PCM focused millimeter-wave sheet beam TWT. In Proceedings of the 2017 Eighteenth International Vacuum Electronics Conference (IVEC), London, UK, 24–26 April 2017; pp. 1–3.

Article

Investigation of a Miniaturized E-Band Cosine-Vane Folded Waveguide Traveling-Wave Tube for Wireless Communication

Kexin Ma, Jun Cai * and Jinjun Feng

National Key Laboratory of Science and Technology on Vacuum Electronics, Beijing Vacuum Electronics Research Institute, Beijing 100015, China; makexin@163.com (K.M.); fengjj@ieee.org (J.F.)
* Correspondence: caijun@sdu.edu.cn; Tel.: +86-130-2615-5686

Abstract: To realize the miniaturization of E-band traveling-wave tubes (TWTs), the size analysis and optimization design were carried out based on an improved cosine-vane folded waveguide (CV-FWG) slow-wave structure (SWS) that operates in a low voltage. In addition, a novel miniaturized T-shaped coupler was proposed to achieve a good voltage standing wave rate (VSWR) in a broad bandwidth. The coupler length was reduced by as much as 77% relative to an original design. With higher coupling impedance, the radius and length of the shortened SWS were optimized as 1.3 mm and 50 mm, respectively. Using microwave tube simulator suit (MTSS) and CST particle studio (PS), 3D beam–wave simulations at 9400 V, 20 mA predicted a gain of 20 dB and a saturated output power of 9 W. The simulation results for CV-FWG TWTs were compared with conventional FWG TWTs from 81 GHz to 86 GHz, showing significant performance advantages with excellent flatness for high-rate wireless communication in the future. The CV-FWG SWS circuit will be fabricated by 3D printing, and this work is underway.

Keywords: TWT; E-band; wireless communication; cosine-vane; FWG; miniaturization

Citation: Ma, K.; Cai, J.; Feng, J. Investigation of a Miniaturized E-Band Cosine-Vane Folded Waveguide Traveling-Wave Tube for Wireless Communication. *Electronics* **2021**, *10*, 3054. https://doi.org/10.3390/electronics10243054

Academic Editor: Alessandro Gabrielli

Received: 10 November 2021
Accepted: 5 December 2021
Published: 7 December 2021

1. Introduction

The Federal Communications Commission (FCC) opened up the E-band (71–76 GHz and 81–86 GHz) for millimeter wave frequency microwave communication [1]. It can meet the demands of high data rate communication with 10 GHz available bandwidth [2]. The transmission rate of E-band can reach 100 Gbit/s, which enables numerous applications, such as local area networks, broadband metropolitan links, back-haul interconnects, and transmissions among next-generation base stations [3–5]. Due to atmospheric and rain attenuation, a high-power amplifier is required to ensure the transmission distance and signal coverage area.

Among vacuum electron devices (VEDs), the traveling-wave tube (TWT) is a preferable choice for millimeter waves, which have high power and wide bandwidth [6–8]. There is much research on E-band TWTs. L-3 Electron Devices division (CA, USA) developed a CW E-band microwave power module (MPM) to cover 81–86 GHz for communication applications. The power amplifier in the MPM is a folded waveguide (FWG) TWT, capable of 80 W saturated output power at 20.8 kV, 220 mA [9]. BVERI developed an E-band TWT that can produce more than 75 W continuous wave saturated output power over the range of 81–86 GHz with a voltage of 16.3 kV and a current of 105 mA [10]. UESTC developed an E-band FWG TWT. The experimental results showed that the prototype tube covers the bandwidth of 83–86 GHz with an output power above 30 W. The tube is tested when the electron gun voltage is 17 kV and the beam current is 62 mA [11]. Both of these TWTs have a large structure size and work at high voltage and high current.

Moreover, the size of the devices is also crucial as TWTs are mounted on the compact platform with a limited space. A 94 GHz, 25 W compact helical TWT for operating at 9700 V, 13.5 mA is being developed with a size of 66.04 mm × 48.26 mm × 48.26 mm [12]. Utilizing the same design rules and fabrication technology, an E-band TWT with a saturated output power of 8 W will be built [13]. Due to the high technical complexity, no subsequent test results have been published so far.

Compared to the helix slow-wave structure (SWS), folded waveguide (FWG) SWS is widely used in millimeter waves. Otherwise, the transverse size of the FWG determines the cutoff frequency of the waveguide, which is difficult to reduce to realize miniaturization. The cosine-vane FWG (CV-FWG) could achieve a significant decrease in the cutoff frequency, which facilitates a transverse miniaturized design of the SWS [14].

This paper is organized as follows. The introduction is presented in Section 1. Section 2 gives the effects of various structure parameters on the cold characteristics. The design of a miniaturized high-frequency structure is described in Section 3, which includes a miniaturized SWS, a miniaturized T-shaped coupler, and the simulation of a beam–wave interaction. Finally, a conclusion is drawn in Section 4.

2. Analysis of Cold Characteristics

Based on normal CV-FWG [14], a modified CV-FWG was studied to add design dimensions by changing the height of the waveguide-connecting section and name it as *d* as shown in Figure 1. The structural parameter *d* makes the structural design more flexible. Figure 1a shows the 3D cut-away of a CV-FWG SWS in a single period, where *a* is the width of the waveguide, *b* is the width of the narrow side, *2p* is the geometric period, *h* is the height of the straight waveguide, r_c is the radius of the electron beam channel, and A_c is the amplitude of the cosine curve whose value could be chosen from zero (conventional FWG) to the sum of *h/2* and *d/2*, and Figure 1b shows a cross-sectional view of the CV-FWG. CV-FWG can be regarded as the combination of FWG and a staggered double vane when the value of *Ac* is greater than *h/4*, the symmetrical profiling in *x*-axis of CV-FWG can be seen as a FWG, and the end view of the CV-FWG can be seen as a staggered double vane. With the increase of *Ac*, the proportion of staggered double vane increases gradually.

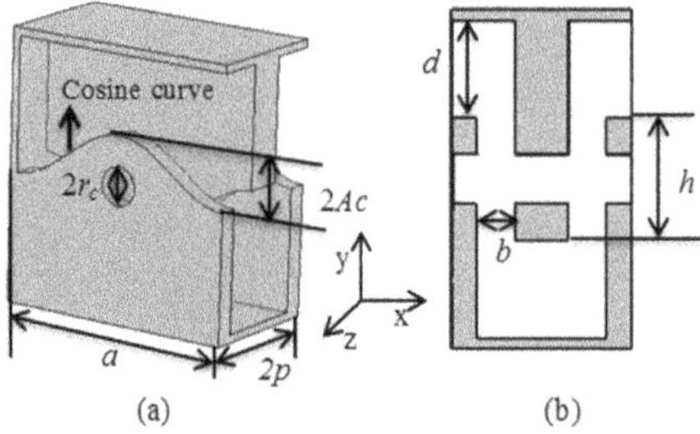

Figure 1. Schematic of single period CV-FWG: (**a**) 3D cut-away solid model and (**b**) cross-sectional view of the CV-FWG.

To realize the miniaturization of SWS, the influences of structure parameters on cold characteristics were analyzed. For the CV-FWG, only the *Ac* and *d* need to be illustrated, the influences of the other structure parameters, including *a*, *b*, *p*, and *h*, are similar to the conventional FWG.

As shown in Figure 2, the normalized phase velocity of CV-FWG decreases with the increase of *Ac*, and the cold bandwidth becomes wider; meanwhile, the cutoff frequency decreases rapidly. When the *Ac* is equal to 0, the SWS is a conventional FWG, and the dispersion is a normal dispersion. As the *Ac* increases, it becomes an abnormal dispersion. Figure 3 shows the effect of *d* on the dispersion. As *d* increases, the dispersion curves become flatter with the reduced intrinsic bandwidths. The normalized phase velocity in the frequency range below 55 GHz increases with the increase of *d*, while the trend reverses above 55 GHz. The cutoff frequency is almost constant.

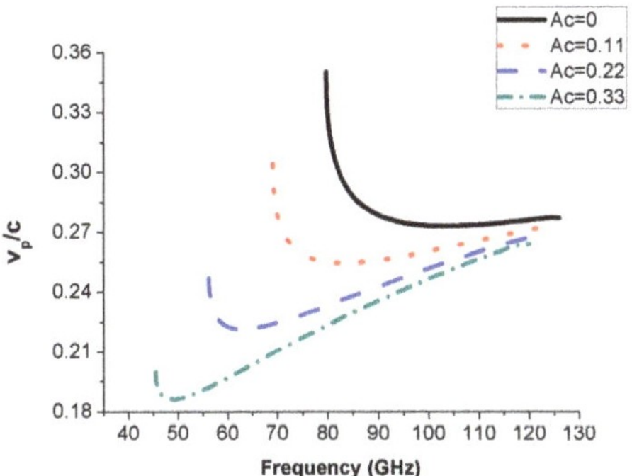

Figure 2. Normalized phase velocity of the CV-FWG for various values of *Ac*.

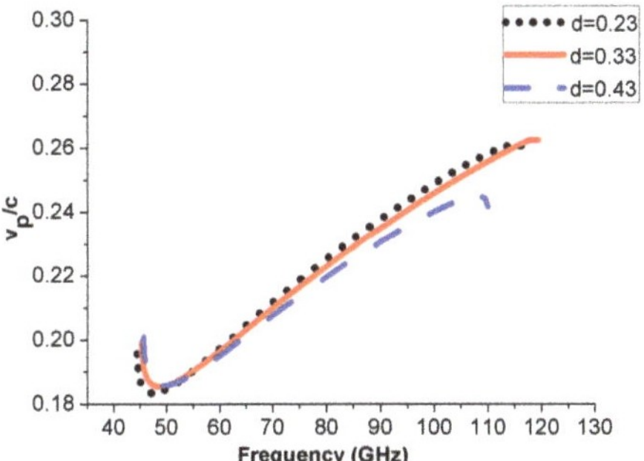

Figure 3. Normalized phase velocity of the CV-FWG for various values of *d*.

Figures 4 and 5 show the effects of the variations of *Ac* and *d* on the on-axis coupling impedance, respectively. At the operation frequency from 81–86 GHz, the on-axis coupling impedance is decreased with the increase of *Ac*, and the curve is flatter, while it is increased with the increase of *d*, and the on-axis coupling impedance flatness are similar to each other.

The loss was calculated with the conductivity of 1.6×10^7 S/m. As shown in Figure 6, with the increase of the *Ac*, the loss is decreased. But the effect of the *d* on the loss is complex as shown in Figure 7. In the frequency range from 81–86 GHz, the variation of loss is within 0.3 dB/m. Though the change is not obvious, when *d* and *h* are adjusted simultaneously, the loss decreases and the coupling impedance increases, while the dispersion is almost unchanged, which is important for the optimized design in the next section.

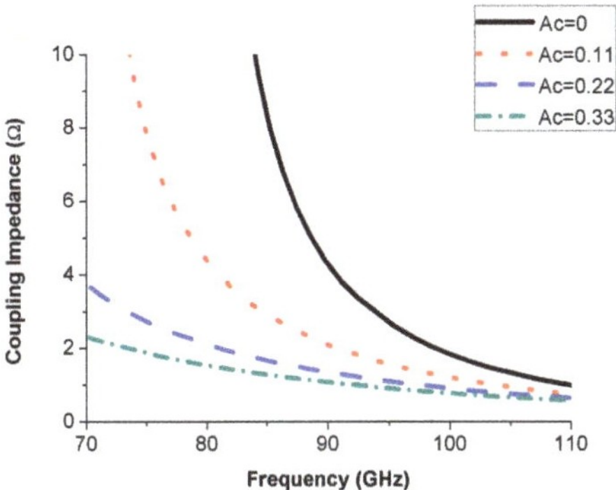

Figure 4. Coupling impedance of the CV-FWG for various values of *Ac*.

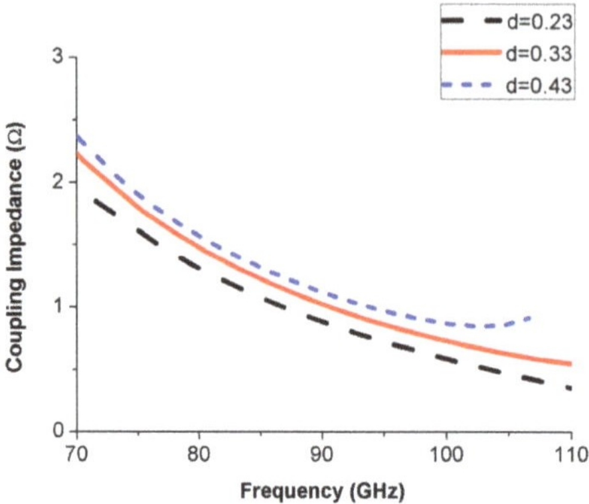

Figure 5. Coupling impedance of the CV-FWG for various values of *d*.

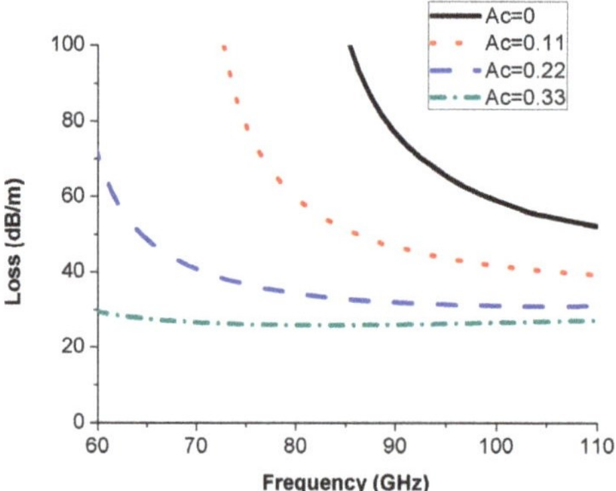

Figure 6. Loss of the CV-FWG for various values of *Ac*.

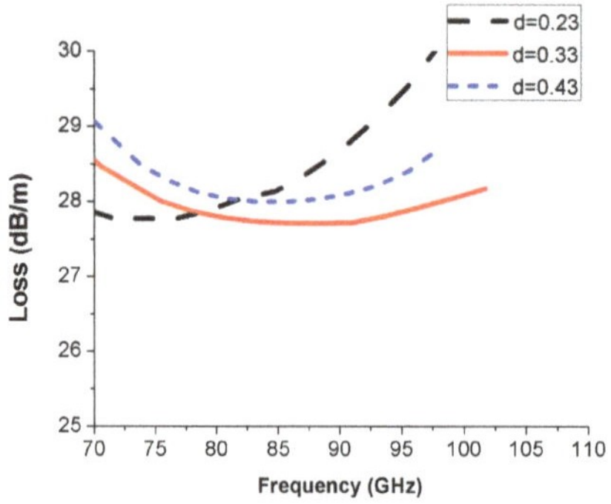

Figure 7. Loss of the CV-FWG for various values of *d*.

3. Design of Device

3.1. Design of SWS

Based on the above analysis, the operational parameters are shown in Table 1. Several SWS designs were investigated to minimize size and make the SWS operate at the designated voltage. The final structure parameters are shown in Table 2. To show the advantages of CV-FWG, a FWG with a similar dispersion in the operation frequency band was designed for comparison, and the structure parameters are also shown in Table 2. Two cross-section structures are compared in Figure 8. The R_1 and R_2 are the radii of the FWG and CV-FWG, respectively. Compared to the FWG, the radius of CV-FWG is reduced by 23.5%, and the area of the cross-section is reduced by 41.8%.

Table 1. Operational parameters.

Frequency	81–86 GHz
Beam voltage	Less than 10 kV
Beam current	Less than 20 mA
Peak output power	More than 8 W
RF input/output	WR-10

Table 2. Structure parameters and comparison.

Parameter	Dimensions (mm)		Reduction Percentage (%)
	CV-FWG	FWG	
a	1.6	2.4	33
b	0.2	0.3	33.33
p	0.45	0.55	18
h	0.6	1	40
r_c	0.12	0.12	0
Ac	0.2	-	-
d	0.5	-	-
R	1.3	1.7	23.5
S (mm^2)	5.3	9.1	41.8

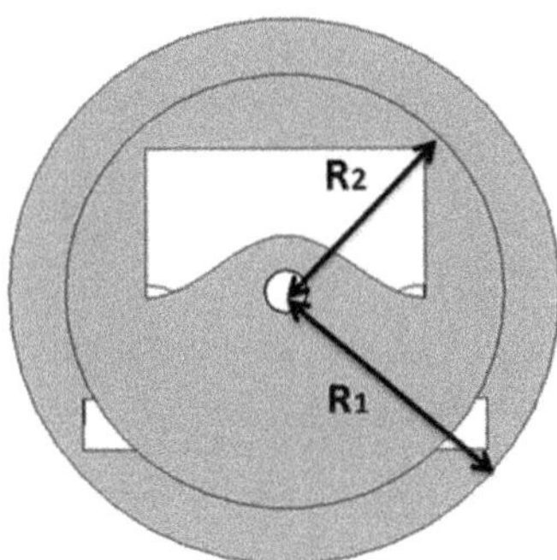

Figure 8. Cross-sectional view of CV-FWG and FWG.

The dispersion curves of CV-FWG and FWG are shown in Figure 9. The dispersions of the two structures are almost identical in the operation frequency from 81–86 GHz, and the curve is very flat with the variation of 0.00025. The loss and the on-axis coupling impedance are compared as shown in Figures 10 and 11. The on-axis coupling impedances for CV-FWG and FWG are 6.5 Ω and 2.1 Ω, respectively; at 81 GHz, the coupling impedance of the CV-FWG is about three times higher than that of the FWG. And at 86 GHz, the coupling impedances are 4.4 Ω and 1.4 Ω, respectively. The coupling impedance of CV-FWG is about 3.2 times higher than that of FWG. Figure 11 shows the losses of the CV-FWG and FWG. Compared with FWG, the loss of CV-FWG is increased by 31.6% at 81 GHz and 35.6% at 86 GHz. It can be seen that, compared with the FWG, the increase of the coupling impedance of CV-FWG is much larger than the increase of the loss. As a result, an increased interaction can be expected.

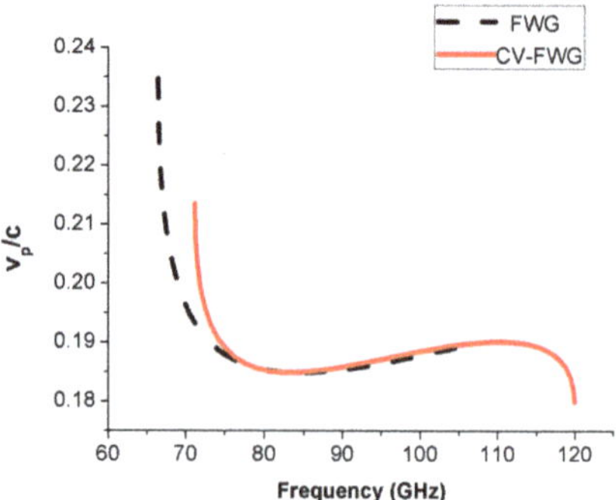

Figure 9. Normalized phase velocity of CV-FWG and conventional FWG.

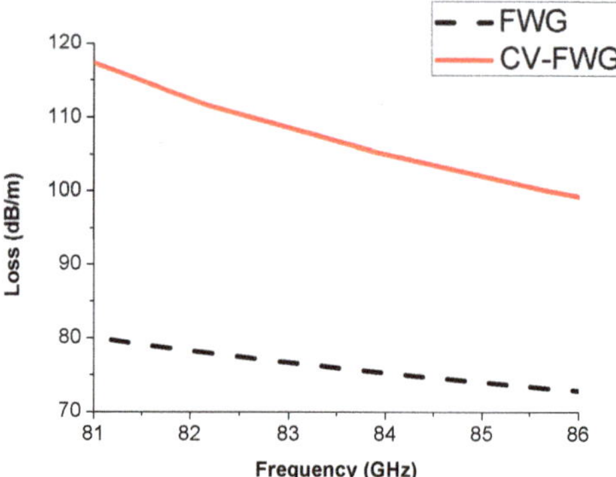

Figure 10. Transmission loss of CV-FWG and conventional FWG.

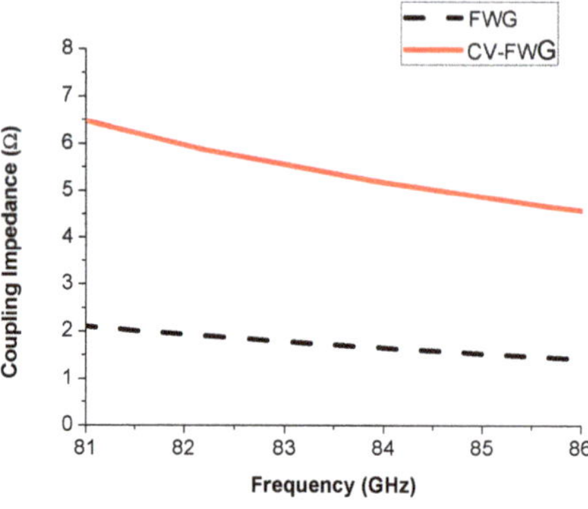

Figure 11. Coupling impedance of CV-FWG and conventional FWG.

It should be pointed out that the on-axis coupling impedance and loss for CV-FWG are smaller than those for FWG in [11], which is different from the above simulation results. The reason is that the comparison is carried out under the same structure parameters in [11]. To obtain consistency and good cold characteristics in the operation frequency band, the operation point of CV-FWG moves to the left, which leads to the increases of loss and on-axis coupling impedance, and the effect of on-axis coupling impedance is more apparent.

3.2. Miniaturized Coupler

It is impossible for CV-FWG to be directly coupled by a WR-10 standard rectangular waveguide. For miniaturization of the tube, a novel coupler named a T-shaped coupler was proposed, as shown in Figure 12.

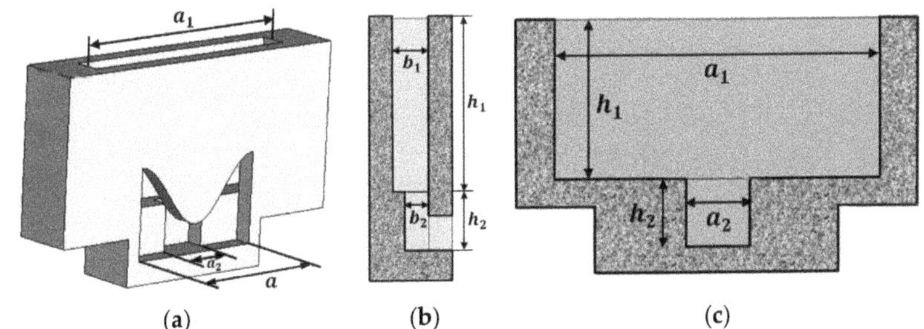

Figure 12. Schematic of the T-coupler: (**a**) 3D cut-away solid model, (**b**) front view of the T-coupler, and (**c**) cross-sectional view of the T-coupler.

The coupler can be regarded as the superposition of two rectangular waveguides with different sizes of length, width, and height. The a_1 and a_2 are the length of the upper rectangular waveguide and the lower rectangular waveguide, respectively. The b_1 and b_2 are the width of the upper rectangular waveguide and the lower rectangular waveguide, respectively. The h_1 and h_2 are the height of the upper rectangular waveguide and the lower rectangular waveguide, respectively.

Using CST Microwave Studio (MWS), the T-shaped coupler was optimized to meet the requirements of the E-band CV-FWG. Considering the difficulty in machining and the roughness of the circuit wall, the conductivity was set to 1.6×10^7 S/m. Figure 13 shows the voltage standing wave ratio (VSWR) as a function of frequency. The VSWR is $\leq$1.25 from 80 GHz to 90 GHz. The miniaturized design was realized because b_1 is equal to 0.3 mm, which is much smaller than the width of a WR-10 standard rectangular waveguide. The coupler length is reduced by as much as 77% relative to an original design.

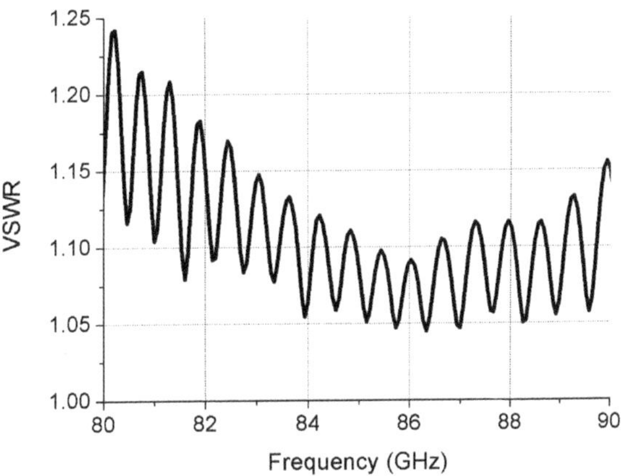

Figure 13. Matching characteristics of the T-shaped coupler.

3.3. Beam–Wave Interaction

The beam–wave interaction circuit was firstly carried out and optimized by microwave tube simulator suit (MTSS) software, then the result was verified by CST Particle Studio

(PS). Figure 14 shows the results of the saturated output power, and the gain for the CV-FWG and FWG were calculated by MTSS with the same operation voltage, current, and interaction length. The saturated output power of CV-FWG is greater than 8 W and the gain is greater than 19 dB, while output power of FWG is less than 2 W and the gain is less than 13 dB. It was proved that the beam–wave interaction of CV-FWG is stronger compared to the FWG and facilitates the reduction of the length of SWS.

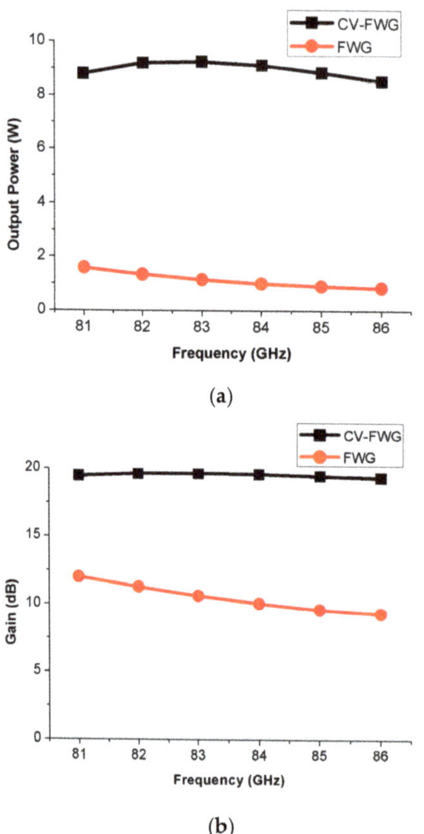

Figure 14. The output power (**a**) and gain (**b**) of CV-FWG and FWG simulated by MTSS.

3D PIC simulation software was used to predict the whole tube performance. A 3D model of a single section CV-FWG was built, which included the input and output couplers and a lossy metal housing (the conductivity was 1.6×10^7 S/m as shown in Figure 15. The PIC simulations were carried out on a cloud computing platform with about seven million mesh cells. Then the beam–wave interaction was calculated with a voltage of 9400 V, a current of 20 mA, and an interaction length of less than 50 mm. The radius of the electron beam is 0.06 mm, which is half of the radius of the electron beam channel. A sinusoidal RF-driven signal was applied. A uniform longitudinal magnetic field of 0.28 T was used in the PIC simulation. The simulation time is 6 ns for the bunching and amplifying. The total number of particles in the model is related to the model size and the simulation time. There are more than 1.73 million particles in this model. Figure 16 shows an amplified output signal at 83 GHz, with saturated power of more than 10 W less than 90 mW. The results of saturated power and gain are shown in Figure 17.

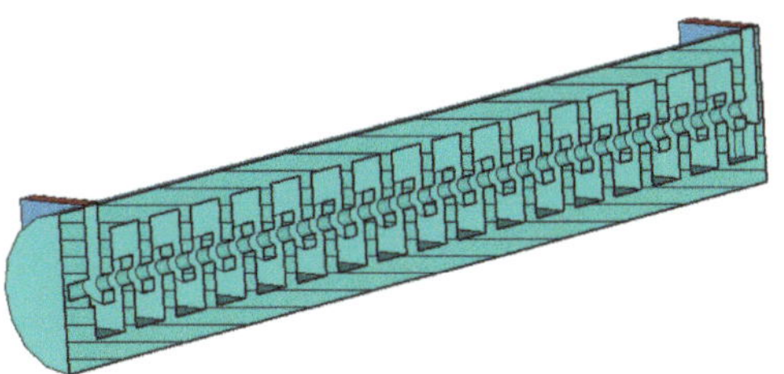

Figure 15. Cross-sectional view of SWS with T-coupler.

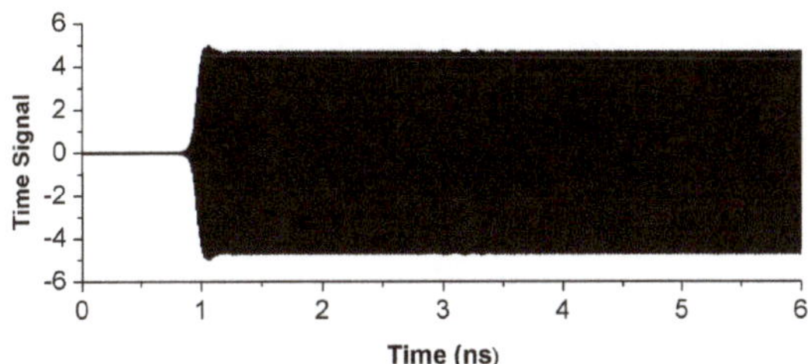

Figure 16. The output signal with 90 mW driven power.

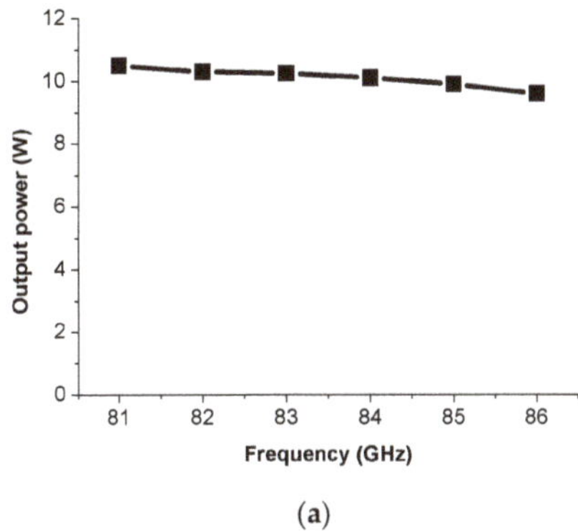

(**a**)

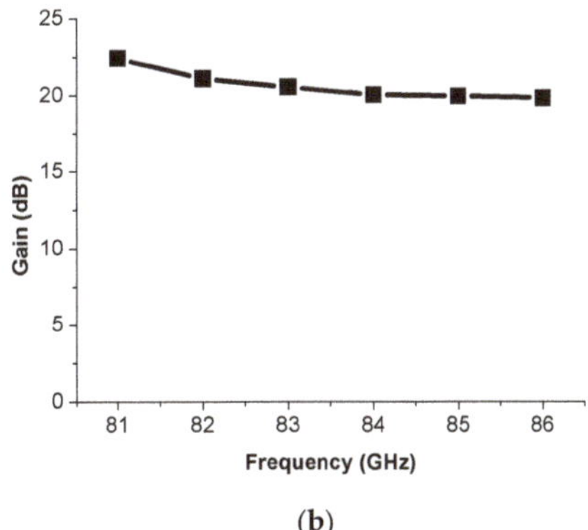

(**b**)

Figure 17. The output power (**a**) and gain (**b**) of CV-FWG simulated by PIC.

It can be seen that the saturated output power of CV-FWG is more than 9 W, and the saturated output power is 10.83 W in 83 GHz. The gain of the CV-FWG is about 20 dB. The gain varies within 20 dB from 81 GHz to 86 GHz. It shows that the gain has great consistency, which means the gain varies very little with frequency. Because the gain is more than 20 dB, the circuit was designed with 0 mW driven and the result indicated the single section slow-wave circuit will not oscillate and the performance of CV-FWG is steady.

4. Conclusions

We introduced a miniaturized CV-FWG SWS that was designed to operate in the voltage <10 kV from 81 GHz to 86 GHz. The dimensions of the CV-FWG SWS are 1.6 mm × 1.6 mm × 47.75 mm. To couple the CV-FWG, a compact T-shaped coupler was proposed with a good matching property. Lastly, the beam–wave interaction was simulated, and the saturated output power is more than 9 W, and the gain has great consistency. This research laid a foundation for the development of follow-up devices, which can be widely used in high data rate wireless communication.

Author Contributions: Conceptualization, K.M. and J.C.; methodology, K.M.; validation, J.C.; formal analysis, K.M.; writing-original draft preparation, K.M.; writing-review and editing, J.F. and J.C.; visualization, K.M.; funding acquisition, J.F. and J.C.; All authors have read and agreed to the published version of the manuscript.

Funding: This research was funded by National Natural Science Foundation of China grant number 61831001.

Conflicts of Interest: The authors declare no conflict of interest.

References

1. Federal Communication Commission: FCC 03-248. *Allocation and Service Rules for the 71-76GHz, 81-86GHz, and 92-95GHz Bands*; Federal Communications Commission: Washington, DC, USA, 2003.
2. Lu, J.; Zhou, Z.; Wang, L.; Zeng, J.; Chen, S. Zero-multiplication TX-filter structure for E-band Transmission system. In Proceedings of the 2014 International Conference on Information and Communications Technologies, Nanjing, China, 16–18 May 2014; pp. 1–5.
3. Ferndahl, M.; Gavell, M.; Abbasi, M.; Zirath, H. Highly integrated E-band direct conversion receiver. In Proceedings of the 2012 IEEE Compound Semiconductor Integrated Circuit Symposium, La Jolla, CA, USA, 14–17 October 2012; pp. 1–4.
4. Ghassemi, N.; Wu, K. Planar High-Gain Dielectric-Loaded Antipodal Linearly Tapered Slot Antenna for E- and W-Band Gigabyte Point-to-Point Wireless Services. *IEEE Trans. Antennas Propag.* **2013**, *61*, 1747–1755. [CrossRef]
5. Ghassemi, N.; Gauthier, J.; Wu, K. Low-Cost E-Band Receiver Front-End Development for Gigabyte Point-to-Point Wireless Communications. In Proceedings of the 43rd European Microwave Conference, Nuremberg, Germany, 6–10 October 2013; pp. 1011–1014.
6. Pan, P.; Zi, Z.; Cai, J.; Tang, Y.; Liu, S.; Xie, Q.; Bian, X.; Feng, J. Millimeter Wave Traveling Tubes for High Data Rate Wireless Communication. *Acta Electronica Sinica* **2020**, *48*, 1834–1840.
7. Li, Y.; Yue, L.; Qiu, B.; Gao, H.; Wang, S.; Zhao, G.; Xu, J.; Yin, H.; Duan, Z.; Huang, M.; et al. Design and Cold Test of a Ka-band Fan-Shaped Metal Loaded Helix Traveling Wave Tube. In Proceedings of the 2020 IEEE 21st International Conference on Vacuum Electronics (IVEC), Monterey, CA, USA, 19–22 October 2020; pp. 241–242.
8. Liu, S.; Xie, Q.; Chen, Z.; Wu, Y.; Zi, Z.; Wu, X.; Cai, J.; Feng, J. High Linear Power E-Band Traveling-Wave Tube for Communication Applications. *IEEE Trans. Electron Devices* **2021**, *68*, 2984–2989. [CrossRef]
9. Kowalczyk, D.; Zubyk, A.; Meadows, C.; Schoemehl, T.; True, T.; Martin, M.; Kirshner, M.; Armstrong, C. High Efficiency E-Band MPM for Communications Applications. In Proceedings of the IEEE 17th International Vacuum Electronics Conference (IVEC), Monterey, CA, USA, 19–21 April 2016; pp. 513–514.
10. Zi, Z.; Liu, S.; Xie, Q.; Li, S.; Cai, J.; Zhao, S. A 70W 81-86GHz E-band CW Traveling Tube. In Proceedings of the IEEE International Vacuum Electronics, Busan, Korea, 28 April–1 May 2019; pp. 1–2.
11. Ji, R.; Yang, Z.; Guo, Z.; Wang, Q.; Han, P.; Gong, H. Design and Experiment of An E-band Folded Waveguide Traveling Wave Tube. In Proceedings of the IEEE 20th International Vacuum Electronics Conference (IVEC), Busan, Korea, 28 April–1 May 2019; pp. 1–2.
12. Kory, C.; Dayton, J.; Mearini, G.; Lueck, M. Microfabricated 94 GHz TWT. In Proceedings of the International Vacuum Electronics, Monterey, CA, USA, 17 July 2014; pp. 175–176.
13. Dayton, J.; Kory, C.; Mearini, G. Microfabricated mm-wave TWT Platform for Wireless Backhaul. In Proceedings of the International Vacuum Electronics, Beijing, China, 27 August 2015; pp. 1–2.
14. Cai, J.; Wu, X.; Feng, J. A Cosine-Shaped Vane-Folded Waveguide and Ridge Waveguide Coupler. *IEEE Trans. Electron Devices* **2016**, *63*, 2544–2549. [CrossRef]

electronics

MDPI

Article

A 340 GHz High-Power Multi-Beam Overmoded Flat-Roofed Sine Waveguide Traveling Wave Tube

Jinjing Luo [1], Jin Xu [1,*], Pengcheng Yin [1], Ruichao Yang [1], Lingna Yue [1,*], Zhanliang Wang [1], Lin Xu [1], Jinjun Feng [2], Wenxin Liu [3] and Yanyu Wei [1]

[1] National Key Laboratory of Science and Technology on Vacuum Electronics, School of Electronic Science and Engineering, University of Electronic Science and Technology of China, Chengdu 610054, China; jinjingluo_uestc@163.com (J.L.); 201811022514@std.uestc.edu.cn (P.Y.); ruichaoyang@foxmail.com (R.Y.); wangzl@uestc.edu.cn (Z.W.); xulin@uestc.edu.cn (L.X.); yywei@uestc.edu.cn (Y.W.)

[2] National Key Laboratory of Science and Technology on Vacuum Electronics, Beijing Vacuum Electronics Research Institute, Beijing 100015, China; fengjinjun@tsinghua.org.cn

[3] Aerospace Information Research Institute, Chinese Academy of Sciences, Beijing 100190, China; lwexin@mail.ie.ac.cn

* Correspondence: alionxj@uestc.edu.cn (J.X.); lnyue@uestc.edu.cn (L.Y.)

Abstract: A phase shift that is caused by the machining errors of independent circuits would greatly affect the efficiency of the power combination in traditional multi-beam structures. In this paper, to reduce the influence of the phase shift and improve the output power, a multi-beam shunted coupling sine waveguide slow wave structure (MBSC-SWG-SWS) has been proposed, and a multi-beam overmoded flat-roofed SWG traveling wave tube (TWT) based on the MBSC-SWG-SWS was designed and analyzed. A TE_{10}-TE_{30} mode convertor was designed as the input/output coupler in this TWT. The results of the 3D particle-in-cell (PIC) simulation with CST software show that more than a 50 W output power can be produced at 342 GHz, and the 3 dB bandwidth is about 13 GHz. Furthermore, the comparison between the single-beam sine waveguide (SWG) TWT and the multi-beam overmoded SWG TWT indicates that the saturated output power of the multi-beam overmoded SWG TWT is three times more than that of the single beam SWG TWT.

Keywords: multi-beam; over-mode; 340 GHz; TWT

Citation: Luo, J.; Xu, J.; Yin, P.; Yang, R.; Yue, L.; Wang, Z.; Xu, L.; Feng, J.; Liu, W.; Wei, Y. A 340 GHz High-Power Multi-Beam Overmoded Flat-Roofed Sine Waveguide Traveling Wave Tube. *Electronics* **2021**, *10*, 3018. https://doi.org/10.3390/electronics10233018

Academic Editors: Yahya M. Meziani and Paolo Baccarelli

Received: 13 October 2021
Accepted: 1 December 2021
Published: 3 December 2021

1. Introduction

Terahertz (THz) technology has a considerable value in medical treatment, device detection, and many other sectors due to its advantages of high penetrability, low photon energy, strong absorption, etc. [1,2]. Vacuum electronic devices (VEDs), especially a travelling wave tube (TWT) that has high output power and broadband [3–6], is a main method to obtain a THz wave. The performance of the TWT is basically determined by the slow wave structure (SWS), and the sine waveguide (SWG) SWS characterized by low reflection and insertion loss has previously been explored as a potential THz amplifier [7–10]. However, the power capacity and the output power decrease significantly as the frequency increases, and only about ten watts of output power can be generated by 340 GHz TWTs [11–14]. Methods for improving the level of output power in the THz band received particular attention from researchers; the power combination technology was employed in TWTs that could distinctly enhance the output power [15,16]. Nevertheless, the inherent problem with this technology is that the machining errors of independent slow wave circuits would cause a phase shift, which would greatly affect the efficiency of the power combination.

To solve the problem, a multi-beam shunted coupling sine waveguide slow wave structure (MBSC-SWG-SWS) in which the energy of each slow wave circuit can be coupled with the other circuits is proposed in this paper, and an overmoded TWT based on such SWS has been studied, including the high frequency characteristics and the beam–wave interaction property. Since the SWS operates in high mode, a TE10-TE30 mode converter has been designed as the input/output coupler. The comparison of output power between the multi-beam overmoded SWG TWT and single beam SWG TWT shows that the presented TWT can obtain three times the output power of the single beam SWG TWT.

2. High Frequency Characteristics

Figure 1 shows the cross-section view of the structure models, in which the period is p, the wide side of the waveguide is a, the height of the beam channel is t and the amplitude of the sine curve is h. The dimensions of the MBSC-SWG-SWS and the single beam SWG SWS working at 340 GHz were confirmed after optimization. As shown in Table 1, to verify the performance of the MBSC-SWG-SWS, the dimensions of these two structures are the same, except that the wide side of the MBSC-SWG-SWS is three times that of the single beam SWG SWS.

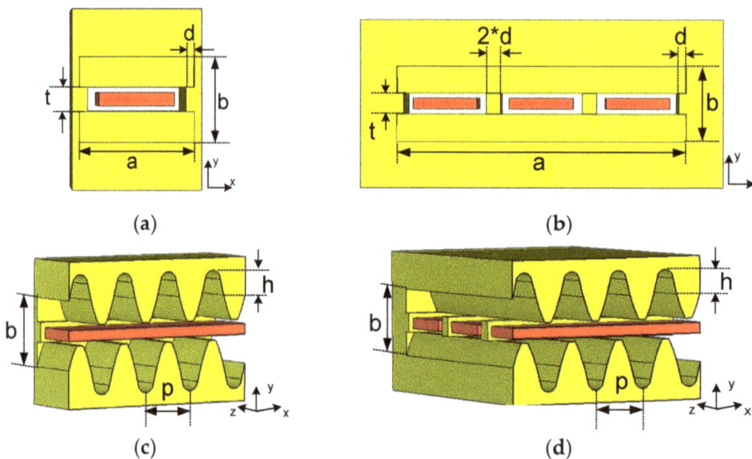

Figure 1. 3D model with sheet beams (red regions). (**a**) Left view of single beam SWG SWS. (**b**) Left view of MBSC-SWG-SWS. (**c**) The cut view of single beam SWG SWS. (**d**) The cut view of MBSC-SWG-SWS.

Table 1. Optimized parameters of MBSC-SWG-SWS and SWG SWS.

Symbol	Value	
	MBSC-SWG-SWS	**SWG SWS**
a	1.605	0.535
b	0.38	0.38
p	0.29	0.29
h	0.15	0.15
d	0.04	0.04
t	0.11	0.11

The dispersion characteristics have been calculated using CST software. As shown in Figure 2, all these three modes have a broadband from 315 to 390 GHz. Figure 3 displays the electric field distribution of the MBSC-SWG-SWS. We found that mode three is the best choice for the operation mode in the MBSC-SWG-SWS because the electric field is evenly distributed in all three tunnels. As shown in Figure 4, the interaction impedance at 340 GHz is 1.1 ohm and the interaction impedance of each tunnel is the same. The results in Figure 5 indicate that both the single beam SWG SWS and the MBSC-SWG-SWS have the same phase velocity in the whole band, which means they have the same synchronous voltage. The normalized phase velocity is 0.281 and the normalized group velocity is 0.197 at 340 GHz.

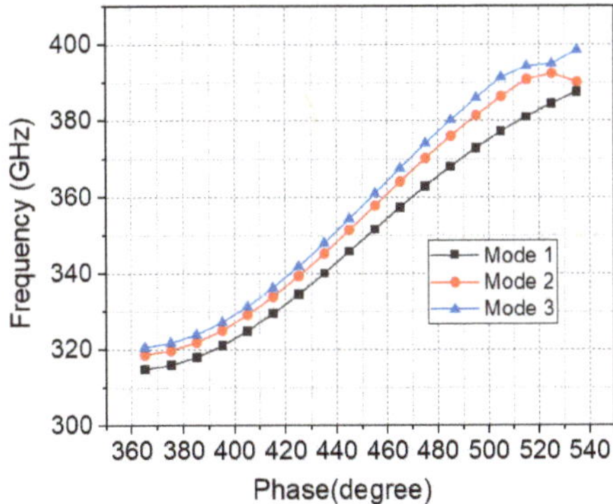

Figure 2. Dispersion curves of MBSC-SWG-SWS.

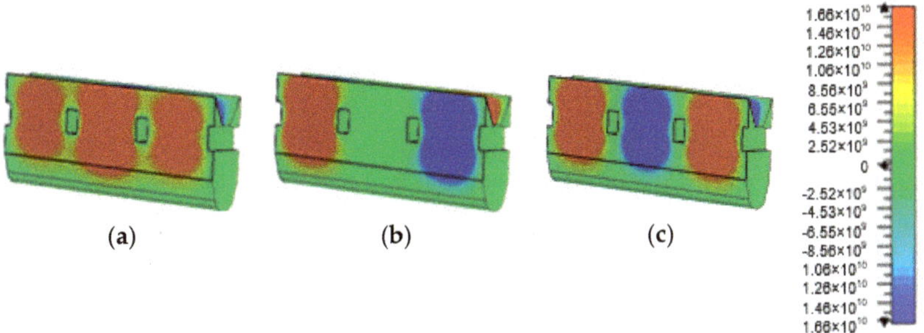

Figure 3. Distribution of electric field along the direction of transmission of MBSC-SWG-SWS. (**a**) Mode 1. (**b**) Mode 2. (**c**) Mode 3.

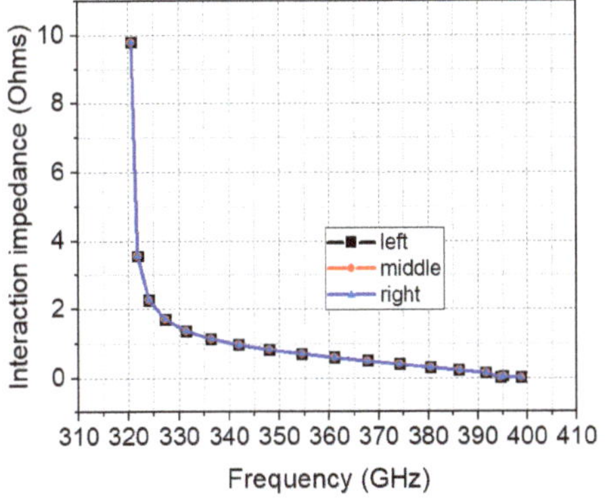

Figure 4. Interaction impedance of three tunnels for mode 3.

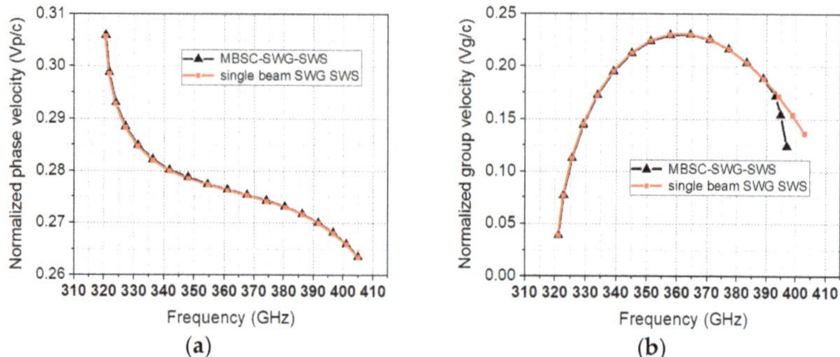

Figure 5. (**a**) Normalized phase velocity versus frequency. (**b**) Normalized group velocity versus frequency.

3. Transmission Characteristics

To transfer the TE_{10} mode in the standard rectangular waveguide WR2 to the TE_{30} mode in the MBSC-SWG-SWS, a TE_{10}-TE_{30} mode converter was designed as the input/output structure in the SWS circuit. Figure 6 shows the vacuum model of the converter and the relevant electric field distribution. Five cylindrical metallic columns were used for mode conversion. The simulation results of the mode convertor are presented in Figure 7. The S_{11} of the mode converter is below −20 dB from 320 to 355 GHz, while the S_{21} is −0.086 dB at 340 GHz and the associated transfer efficiency is 98%.

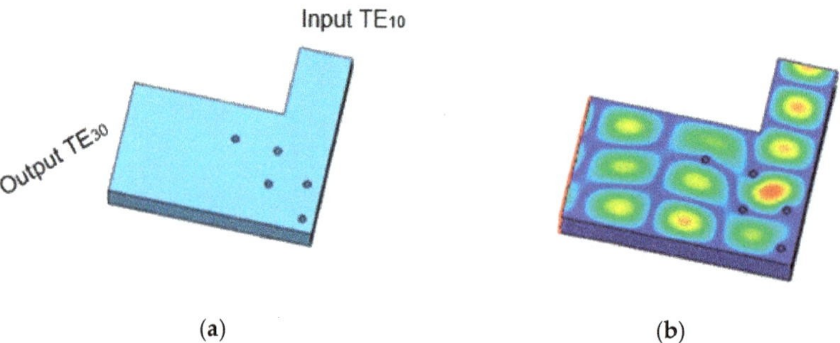

Figure 6. (**a**) Vacuum model of TE_{10}-TE_{30} convertor. (**b**) Electric field of TE_{10}-TE_{30} convertor.

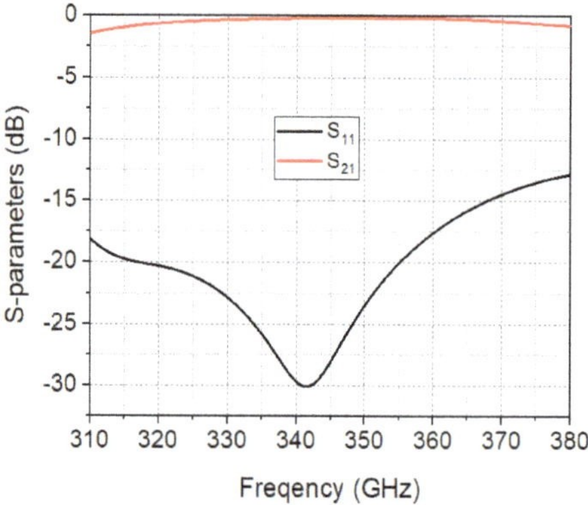

Figure 7. Simulation results of S-parameters for the TE_{10}-TE_{30} convertor.

As shown in Figure 8, the multi-beam overmoded SWG TWT circuit model, including the SWS circuit, electron beam tunnels and mode converters, was built in the CST STUDIO SUITE. The period number of the main SWS circuit is 120. The equivalent conductivity σ_0 can be calculated from the following equations:

$$\sigma_0 = \frac{\sigma}{\left(1 + \frac{2}{\pi}\arctan\left(1.4\left(\frac{R}{\delta}\right)^2\right)\right)^2} \tag{1}$$

$$\delta = \sqrt{\frac{2}{\omega\mu\sigma}} \tag{2}$$

where R is the surface roughness and σ is the conductivity of high conductivity oxygen-free copper. As the surface roughness of the nano-computer numerical control machined model is about 100 nm [17], the effective conductivity is set to 2×10^7 s/m. The transmission characteristic is exhibited in Figure 9. The simulation results show that the S_{11} of the MBSC-SWG-SWS is below –20 dB ranging from 330 to 350 GHz and the S_{21} is -13.5 dB at 340 GHz, while the S_{21} of the single beam SWG SWS is -16.3 dB. We found that the loss of the MBSC-SWG-SWS is smaller than that of the single-beam SWG SWS, which is due to the fact that the MBSC-SWG-SWS has a lower power density than the single beam SWG SWS.

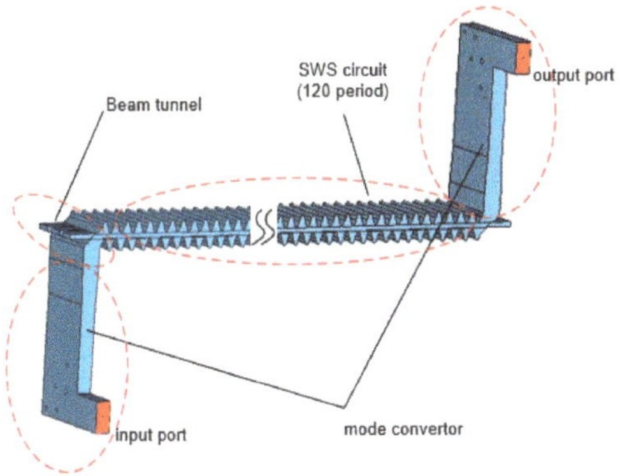

Figure 8. Vacuum model of TWT circuit using MBSC-SWG-SWS.

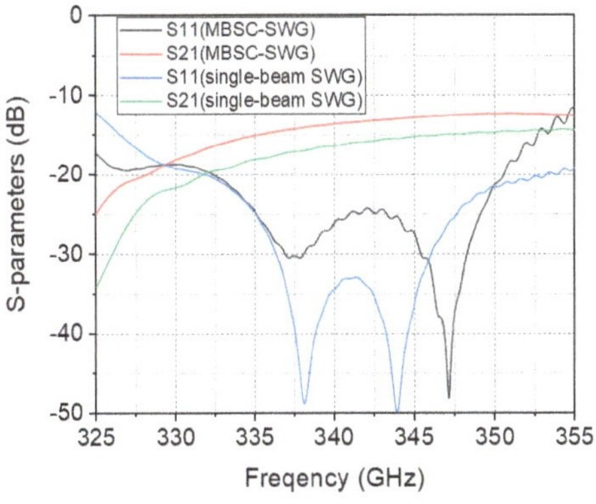

Figure 9. Simulation results of S-parameters for MBSC-SWG-SWS and single-beam SWG SWS.

4. Beam–Wave Interaction

The beam–wave interaction process for the multi-beam overmoded SWG TWT and the single beam SWG TWT were simulated and are compared in this section. Figure 10 shows the models that consisted of 120 periods. According to the aforementioned Brillouin curve of the MBSC-SWG-SWS, the synchronous voltage is 21.3 kV. To study the output power and gain of the multi-beam overmoded SWG TWT, input signals with frequencies between 332 and 350 GHz, an input power of 0.2 W and an operating current of 54 mA were injected in the waveguide port of the amplifier. A uniform magnetic field of 0.7 T was used to focus the electron beams. Researchers have reported studies of the multi-beam electron optics system that indicate that the generation and the focusing of electron beams can be realized [18,19]. The cross-sectional size of each sheet beam was set to 0.3 mm × 0.06 mm, and the filling ratio was 35.9%. The number of mesh cells was set at 29,096,144. The corresponding maximum and minimum mesh steps were 0.017 and 0.008 mm, respectively. The number of macro particles was fixed at 150, and the time step was 0.003 ns.

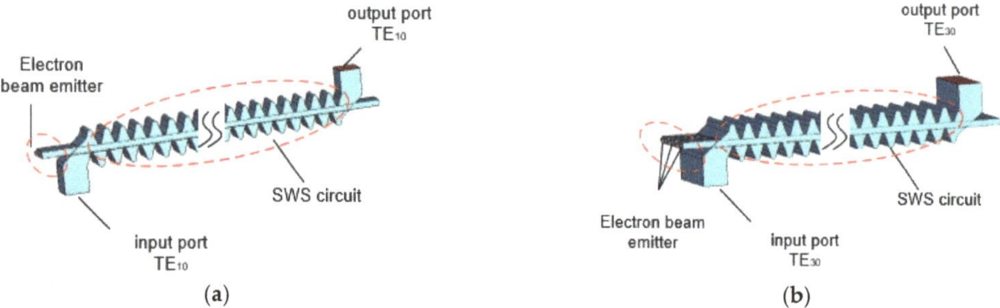

Figure 10. Vacuum model for PIC simulation calculation. (**a**) Single-beam SWG TWT. (**b**) Multi-beam overmoded SWG TWT.

Figure 11 depicts the output power and gain of the multi-beam overmoded SWG TWT versus the operating frequency, respectively. We found that more than 30 W of the output power can be produced from 334 to 347 GHz. The maximum output power is 50 W at 342 GHz and the corresponding gain is 24 dB. Figure 12a gives the time-domain simulation results of the output signal of the multi-beam TWT at 340 GHz, which indicates that the signal is stable after 0.8 ns with a voltage amplitude of 10 V. The potential reflected wave and higher harmonic wave are effectively suppressed, as demonstrated in Figure 12b. The electron bunching effect at the end of the circuit is presented in Figure 13, which illustrates an effective interaction between the electron beams and the electromagnetic wave.

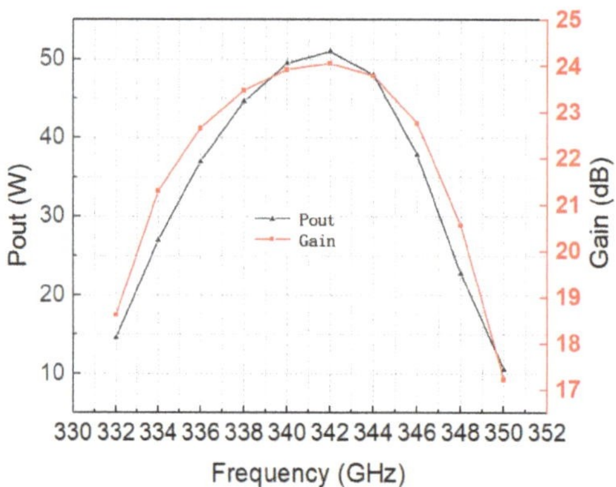

Figure 11. The output power and gain of multi-beam overmoded SWG TWT versus the operating frequency.

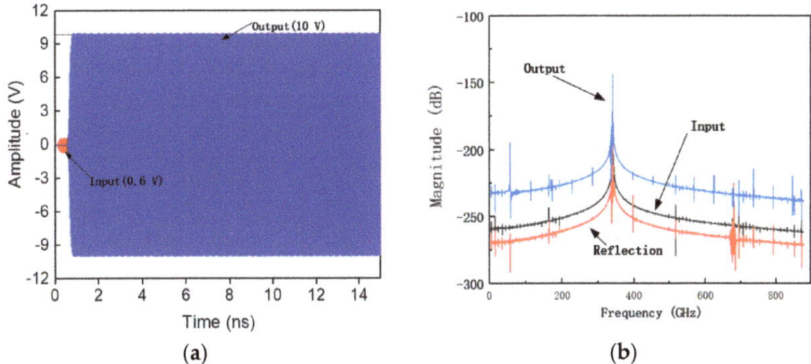

Figure 12. (**a**) Input, output and reflection signals for 340 GHz. (**a**) Time domain. (**b**) Frequency spectrum.

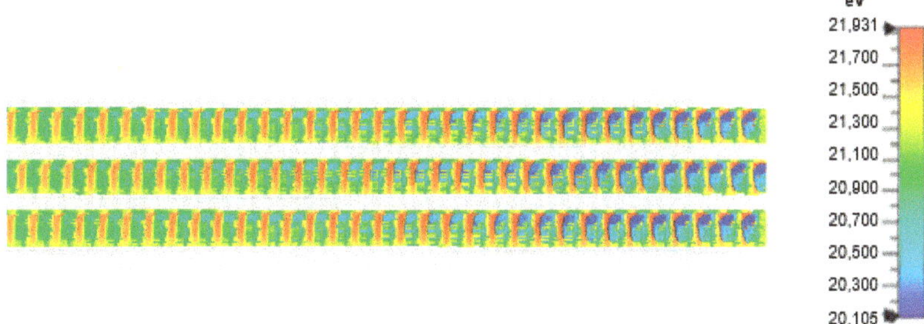

Figure 13. Electrons bunching effect at the end of the circuit.

Figure 14a shows the output power versus input power with the operating current of 36 mA (current density is 200 A/cm^2) at 340 GHz. For the single beam SWG TWT, the saturated output power can reach 10 W when the input power is 0.1 W. Nevertheless, the saturated output power of the multi-beam overmoded SWG TWT is 30 W when the input power is 0.3 W. If the operating current is 54 mA (current density is 300 A/cm^2), as shown in Figure 14b, the saturated output power of the single beam SWG TWT can reach 17 W when the input power is 0.2 W, and the saturated output power of the multi-beam overmoded SWG TWT is 50 W when the input power is 0.3 W. We found that the output power of the multi-beam overmoded SWG TWT is three times that of the single beam SWG TWT.

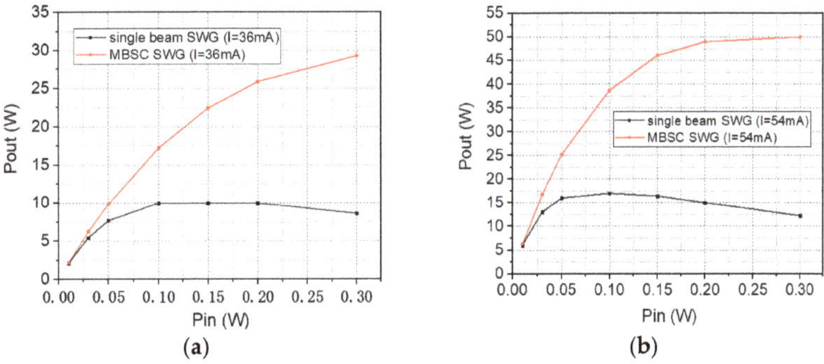

Figure 14. The output power versus the input power at 340 GHz. (**a**) I = 36 mA. (**b**) I = 54 mA.

5. Conclusions

In this paper, an MBSC-SWG-SWS has been proposed to reduce the effect of the phase shift caused by machining errors of independent slow wave circuits, because the energy of each slow wave circuit can be coupled with that of the other circuits. We found that the SWS works in the high order mode through the analysis of its high frequency characteristics; therefore, a TE_{10}-TE_{30} mode convertor has been designed as the input/output couplers. The study of a multi-beam overmoded SWG TWT based on the MBSC-SWG-SWS shows that an output power of more than 30 W can be obtained in the frequency range from 334 to 346 GHz, and the maximum power and the corresponding gain are 51 W and 24 dB at 342 GHz, respectively. Compared with the single beam SWG TWT, the output power of the multi-beam overmoded TWT is three times that of the single beam SWG TWT. These results suggest that the MBSC-SWG-SWS has the ability to enhance the output power. Consequently, the SWS is a potential and promising structure for the high-power THz traveling wave amplifier.

Author Contributions: Conceptualization, J.L., J.X. and Y.W.; methodology, J.L. and Y.W.; software, J.L. and P.Y.; validation, J.L., J.X. and Y.W.; formal analysis, R.Y., L.Y. and L.X.; investigation, R.Y., Z.W. and W.L.; resources, J.L.; data curation, J.L.; writing—original draft preparation, J.L.; writing—review and editing, J.L., J.X., W.L. and J.F. All authors have read and agreed to the published version of the manuscript.

Funding: This work is support by the National key research and development program of China (Grant No. 2017YFE0130000) and the National Natural Science Foundation of China (Grant No. 61771117).

Conflicts of Interest: The authors declare no conflict of interest.

References

1. Siegel, P.H. Terahertz technology. *IEEE Trans. Microw. Theory Tech.* **2002**, *50*, 910–928. [CrossRef]
2. Sherwin, M. Applied physics: Terahertz power. *Nature* **2002**, *420*, 131–133. [CrossRef] [PubMed]
3. Booske, J.H.; Dobbs, R.J.; Joye, C.D.; Kory, C.L.; Neil, G.R.; Park, G.-S.; Park, J.; Temkin, R. Vacuum Electronic High Power Terahertz Sources. *IEEE Trans. Terahertz Sci. Technol.* **2011**, *1*, 54–75. [CrossRef]
4. Booske, J.H. Plasma physics and related challenges of millimeterwave-to-terahertz and high power microwave generation. *Phys. Plasmas* **2008**, *15*, 055502. [CrossRef]
5. Ding, Y.; Liu, P.; Zhang, Z.; Wang, Y. An Overview of Advances in Vacuum Electronics in China. In Proceedings of the 2011 IEEE International Vacuum Electronics Conference (IVEC), Bangalore, India, 21–24 February 2011.
6. Minenna, D.F.; André, F.; Elskens, Y.; Auboin, J.F.; Doveil, F.; Puech, J.; Duverdier, É. The traveling-wave tube in the history of telecommunication. *Eur. Phys. J. H* **2019**, *44*, 1–36. [CrossRef]
7. Xu, X.; Wei, Y.; Shen, F.; Duan, Z.; Gong, Y.; Yin, H.; Wang, W. Sine Waveguide for 0.22-THz Traveling-Wave Tube. *IEEE Electron Device Lett.* **2011**, *32*, 1152–1154. [CrossRef]
8. Xie, W.-Q.; Wang, Z.-C.; He, F.; Luo, J.-R.; Liu, Q.-L. Linear analysis of a 0.22THz sine waveguide travelling wave tube. In Proceedings of the 2014 39th International Conference on Infrared, Millimeter, and Terahertz waves (IRMMW-THz), Tucson, AZ, USA, 14–19 September 2014.
9. Fang, S.; Xu, J.; Yin, H.; Lei, X.; Jiang, X.; Yin, P.; Wu, G.; Yang, R.; Li, Q.; Guo, G.; et al. Experimental Verification of the Low Transmission Loss of a Flat-Roofed Sine Waveguide Slow-Wave Structure. *IEEE Electron Device Lett.* **2019**, *40*, 808–811. [CrossRef]
10. Yang, R.; Xu, J.; Yin, P.; Wu, G.; Fang, S.; Jiang, X.; Luo, J.; Yue, L.; Yin, H.; Zhao, G.; et al. Study on 1-THz Sine Waveguide Traveling-Wave Tube. *IEEE Trans. Electron Devices* **2021**, *68*, 2509–2514. [CrossRef]
11. Cai, K.; Yang, J.; Deng, G.; Yin, Z. Design of 340 GHz Traveling Wave Tube Amplifier Based on Rectangular Staggered Double Vane Slow-Wave Structure. In Proceedings of the 2018 International Conference on Microwave and Millimeter Wave Technology (ICMMT), Chengdu, China, 6–9 May 2018.
12. Hu, P.; Lei, W.; Jiang, Y.; Huang, Y.; Song, R.; Chen, H.; Dong, Y. Development of a 0.32-THz Folded Waveguide Traveling Wave Tube. *IEEE Trans. Electron Devices* **2018**, *65*, 2164–2169. [CrossRef]
13. Choi, W.; Lee, I.; Choi, E. Design and Fabrication of a 300 GHz Modified Sine Waveguide Traveling-Wave Tube Using a Nanocomputer Numerical Control Machine. *IEEE Trans. Electron Devices* **2017**, *64*, 2955–2962. [CrossRef]
14. Wei, S.; Han-Wen, T.; Zhan-Liang, W.; Tao, T.; Hua-Rong, G.; Zhao-Yun, D.; Yan-Yu, W.; Jin-Jun, F.; Yu-Bin, G. Simulation and cold test of a 340 GHz filleted staggered double vane traveling wave tube. *J. Infrared Nfrared Millim. Waves* **2019**, *38*, 303–309.
15. Sheng, L.G.Y.L.M.; Gang, L. Multi-beam TWT with active power combining. *Int. J. Electron.* **1998**, *84*, 647–657.
16. Yan, S.; Su, W.; Xu, A.; Wang, Y. Analysis of multi-beam folded waveguide traveling-wave tube for terahertz radiation. *J. Electromagn. Waves Appl.* **2014**, *29*, 436–447. [CrossRef]

17. Fang, S.; Xu, J.; Hairong, Y.; Yin, P.; Lei, X.; Wu, G.; Yang, R.; Luo, J.; Yue, L.; Zhao, G.; et al. Design and Cold Test of Flat-Roofed Sine Waveguide Circuit for W-Band Traveling-Wave Tube. *IEEE Trans. Plasma Sci.* **2020**, *48*, 4021–4028. [CrossRef]
18. Ruan, C.; Wang, P.; Zhang, H.; Su, Y.; Dai, J.; Ding, Y.; Zhang, Z. Design of planar distributed three beam electron gun with narrow beam separation for W band staggered double vane TWT. *Sci. Rep.* **2021**, *11*, 1–12. [CrossRef] [PubMed]
19. Liang, H.; Xue, Q.; Ruan, C.; Feng, J.; Wang, S.; Liu, X.; Zhang, Z. Integrated Planar Three-Beam Electron Optics System for 220-GHz Folded Waveguide TWT. *IEEE Trans. Electron Devices* **2017**, *65*, 270–276. [CrossRef]

MDPI

Article

Broadband and Integratable 2 × 2 TWT Amplifier Unit for Millimeter Wave Phased Array Radar

Guo Guo [1], Zhenlin Yan [2], Zhenzhen Sun [1], Jianwei Liu [1,*], Ruichao Yang [1], Yubin Gong [1] and Yanyu Wei [1]

[1] National Key Laboratory of Science and Technology on Vacuum Electronics, School of Electronic Science and Engineering, University of Electronic Science and Technology of China, Chengdu 610054, China; guoguo@uestc.edu.cn (G.G.); 18738907505@163.com (Z.S.); ruichaoyang@foxmail.com (R.Y.); ybgong@uestc.edu.cn (Y.G.); yywei@uestc.edu.cn (Y.W.)

[2] Science and Technology on Electronic Information Control Laboratory, Chengdu 610036, China; yanzhenlin29@163.com

* Correspondence: jianwei@uestc.edu.cn

Abstract: A novel power amplifier unit for a phased array radar with 2 × 2 output ports for a vacuum electron device is proposed. Double parallel connecting microstrip meander-lines are employed as the slow-wave circuits of a large power traveling wave tube operate in a Ka-band. The high frequency characteristics, the transmission characteristics, and the beam–wave interaction processes for this amplifier are simulated and optimized. For each output port of one channel, the simulation results reveal that the output power, saturated gain, and 3-dB bandwidth can reach 566 W, 27.5 dB, and 7 GHz, respectively. Additionally, the amplified signals of four output ports have favorable phase congruency. After fabrication and assembly, transmission tests for the 80-period model are performed preliminarily. The tested "cold" S-parameters match well with the simulated values. This type of integratable amplifier combined with a vacuum device has broad application prospects in the field of high power and broad bandwidth on a millimeter wave phased array radar.

Keywords: 2 × 2 amplifier unit; traveling wave tube (TWT); phase congruency; phased array radar (PAR); experimental test

Citation: Guo, G.; Yan, Z.; Sun, Z.; Liu, J.; Yang, R.; Gong, Y.; Wei, Y. Broadband and Integratable 2 × 2 TWT Amplifier Unit for Millimeter Wave Phased Array Radar. *Electronics* **2021**, *10*, 2808. https://doi.org/10.3390/electronics10222808

Academic Editor: Paolo Colantonio

Received: 8 October 2021
Accepted: 12 November 2021
Published: 16 November 2021

1. Introduction

Phased array radar (PAR) has displayed great advantages to users in recent years [1,2]. As the most critical components, transmit/receive (T/R) modules occupy about 70% percent of the PAR antennas. As the terminal amplification component in the transmit link of the T/R modules, the power amplifier (PA) always determines the transmit distance of the PAR. A lightweight, low-voltage, compact and broadband PA with large power at millimeter-wave frequencies (30~100 GHz) is probably satisfied by the requirements of the T/R modules in PAR [3,4].

Traditionally, solid-state semiconductors, such as GaAs and even GaN PA, are widely used in the design of T/R modules [5,6]. For instance, the GaN PA in Ka-band is capable of tens of watts of output power with a bandwidth up to 2 GHz, which could indicate potential improvements both on power and bandwidth. In recent years, vacuum electronic devices have developed rapidly and show great potential at the millimeter wave frequency range and even on the THz spectrum. Both with high power output and broad bandwidth, the traveling-wave tube (TWT) has been widely used in electronic counters, transmitters, and communications [7].

Due to a relatively larger volume and incompatibility to solid-state circuits, TWTs are always placed separately for each transmit channel in a large power active PAR. This traditional framework is not only expensive, but also heavy and bulky, so it has many limitations for application. As we have developed micro-fabrication technologies, the production requirements for novel slow-wave structures (SWSs) are probably met in the millimeter-wave frequency range [8,9]. Among these SWSs, the microstrip meander-line (MML) has been proven to be suitable for low-voltage and broadband TWTs [10,11]. It is also naturally easy to integrate with solid-state circuits and has superiority over the volume.

Motivated by the applications of PARs, a novel broadband 2 × 2 amplifier unit with a large power TWT is proposed in this paper. This novel framework has potential for applying the TWT to the terminal PA in the transmit link of the T/R modules. The simulation results show that these amplifier units are capable of delivering kW-class output power fully with a 3-dB bandwidth of more than 7 GHz in the Ka-band. Moreover, the simulation results show that the phase congruency at the four out ports is great.

The remainder of this paper is organized as follows. Section 2 presents the system model, which includes the novel SWS design and the whole input coupling structure. Simulation results are presented in Section 3. In Section 4, we present our fabrication and experimental tests, and compare the results with the simulation. The final conclusions are presented in Section 5.

2. Structure Description

Easy to be integrated with a solid-state circuit, the MML SWS can be printed with micro fabrication technologies. Of all the modified and improved MML SWSs, the symmetrical double V-shaped MML with higher electron efficiency has been investigated in detail in reference [12]. The schematic of the symmetric double V-shaped MML is shown in Figure 1.

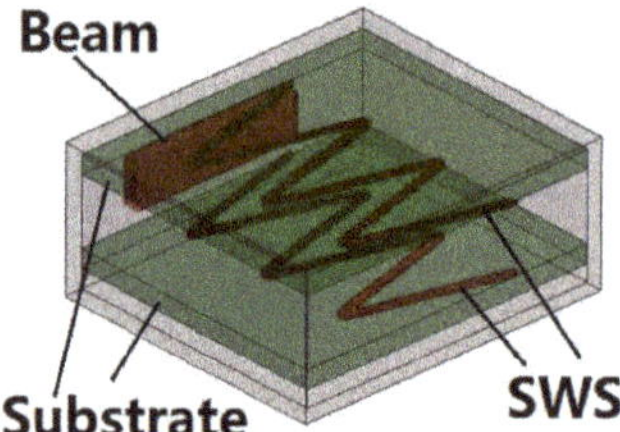

Figure 1. Sketch of the symmetric double V-shape MML SWS.

Inspired by the above TWTs, we propose a novel SWS by parallel, connecting two MMLs together, and then symmetrical, placing them on both the top and bottom dielectric substrates of the cavity to form the four output ports. The two parallel MMLs are connected to each other by power dividers at the end and the whole input coupling structure is designed as a double microstrips–rectangular waveguide transition. Four coaxial cable output ports are connected to the SWS with microstrip lines. Two tapered attenuators are symmetrically inserted into the SWS to suppress the oscillations. The sketch of the novel SWS is shown in Figure 2, in which this 2 × 2 unit consists of one rectangular waveguide input port and four coaxial output ports. One sheet electron beam gun is employed in the beam–wave interaction processes of the TWT to further reduce the whole volume and weight.

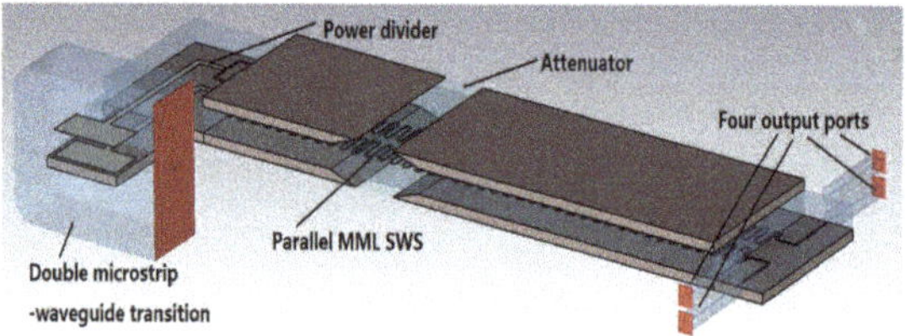

Figure 2. System schematic of the novel 2 × 2 amplifier units.

The operation principle of the amplifier is based on typical O-type vacuum electronic devices. To achieve multi-output, the double microstrips–rectangular waveguide transition and the power divider are employed in the structure. The electron beam is emitted by the electron gun and focused by the focus magnetic field, and then enters into the interaction space. Meanwhile, the input signal is imported into the top and bottom MMLs through the double microstrips and rectangular waveguide transitions, respectively. Then, the signals on the top and bottom MMLs are both divided into two signals by the power divider and transmitted along the four parallel MML SWSs. The four independent signals stimulate radio-frequency fields around the MML SWSs and interact with electron beams. As most of the electrons are decelerated, some kinetic energy from the sheet electron beam is transferred to the radio-frequency fields and the four signals are amplified respectively. At last, the 2 × 2 amplified signals are exported by the four coaxial cable output ports.

This 2 × 2 amplifier unit is probably integrated for larger scale units with multiple methods for application in the T/R modules of the PARs. Based on the development of the multi-beam electron gun [13], one possible framework to compose 2 × 2n units is to simply pile up the units vertically as in Figure 3, in which n is the quantity of the 2 × 2 amplifier unit. In this integrated framework, the input ports can be designed as overall input ports, or even directly connect to the preamp circuit by microstrip lines.

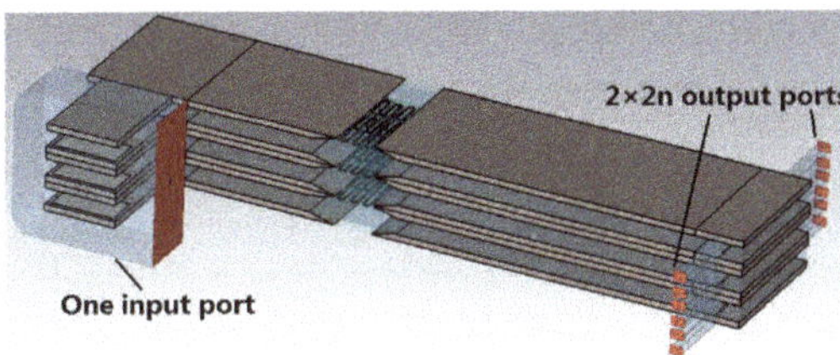

Figure 3. A possible framework for integrated 2 × 2n units.

3. Simulation Results

The high frequency characteristics, including dispersion curves and average interaction impedances for a single period, are first investigated. It can be seen in Figure 4a that the dispersion curves are flat in the frequency between 20 GHz and 40 GHz, which indicates that the cold bandwidth is relatively broad. The averaged interaction impedances (Kc) over the cross section of the electron beam can accurately predict the beam–wave interaction efficiency. According to the transverse parameters of the SWS, the cross section of the electron beam is set to be 2.12 mm × 0.3 mm. The average coupling impedances can reach 2.5 Ohms in the operation frequency band.

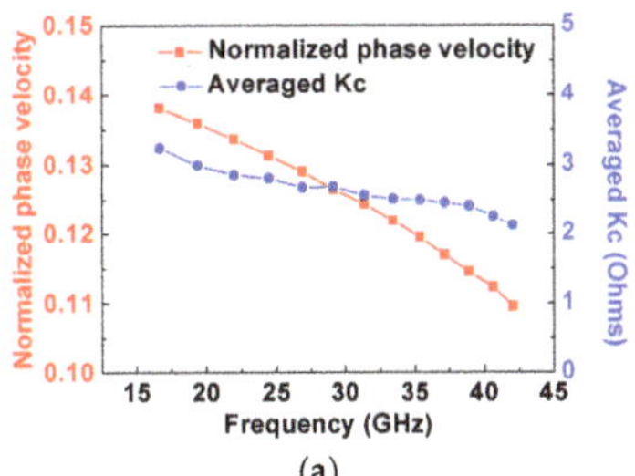

(a)

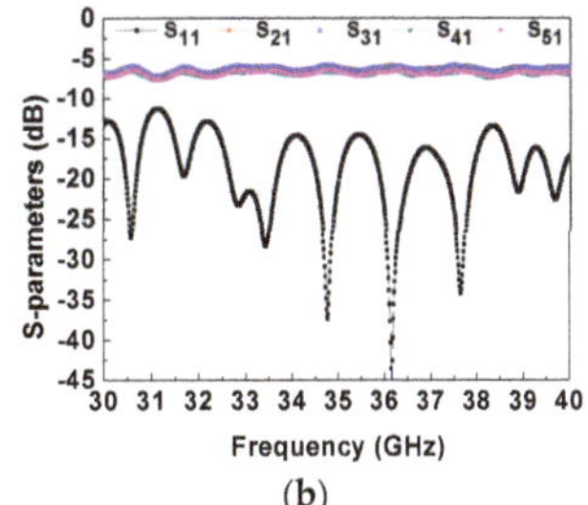

(b)

Figure 4. Plot of the dispersion curve (red) and the averaged interaction impedances (blue) versus frequency (**a**); S-parameters of the transmission model (**b**).

The transmission model without an attenuator is optimized by CST Microwave Studio [14], including 80-period SWS, the input and output coupling structures, and the power combiners, to reduce reflection and improve transmission efficiency. The material of the dielectric substrates is set to Rogers RT5880, with a relative permittivity and dielectric dissipation factor of 2.2 and 0.0009, respectively. The dimensional parameters of the SWS in the simulations are listed in Table 1.

Table 1. Dimensional parameters of the SWS.

Parameter	Value (mm)
Dielectric substrate thickness	0.2
Dielectric substrate width	4.16
Distance between up and down substrates	0.5
MML SWS thickness	0.005
MML SWS transverse width	2
Single pitch length	0.38
Beam cross section	2.12×0.3

The S_{11} parameters are generally below -16 dB and between 33 GHz and 38 GHz, as shown in Figure 4b. The almost coincided S_{21}, S_{31}, S_{41} and S_{51} curves with amplitudes larger than -7 dB indicate symmetry and uniformity for the four output ports.

In order to verify the beam–wave interaction of the novel structure, the whole amplifier model of the novel TWT shown in Figure 2 is built by CST. Two beryllia attenuators are symmetrically inserted into the SWS to suppress the oscillations. The attenuators are optimally designed as a tapered shape to absorb reflected waves over a wide range of frequencies. Driven by only one sheet electron beam with large width–thickness ratio, the beam–wave interaction processes of the novel TWT amplifier are investigated.

More than 3.5 million particles and 8.3 million meshes are used in the particle-in-cell (PIC) simulation. The electrical parameters of the TWT shown in Table 2 have already been optimized by utilizing CST Particle Studio software. The beam voltage, current, and the focus magnetic field in the TWT simulation are set to 5100 V, 1 A, and 0.8 T, respectively. Here we specify the physical meaning of each variables in CST Particle Studio as follows [14]: the sheet beam voltage indicates the kinetic start energy of the emitted particles; the sheet beam current indicates the emission current; the input signal amplitude indicates the amplitude of the input sine step signal; the focus magnetic field indicates the amplitude of a constant focus magnetic field inside the whole computational domain; the center frequency indicates the designed center operation frequency of the amplifier.

Table 2. Optimized electrical parameters for the TWT.

Parameter	Value
Sheet Beam voltage	5100 V
Sheet Beam current	1 A
Input signal amplitude	1 V
Focus magnetic field	0.8 T
Center frequency	35 GHz

The PIC simulation results at a frequency of 35 GHz are shown Figures 5–9. Figure 5 depicts the electron bunching phenomenon in a TWT operation, where we can see that the accelerating electrons and retarding electrons are periodically arranged along a longitudinal direction, which demonstrates a good beam–wave energy exchange process. This physical phenomenon is very typical in O-type vacuum electronic devices.

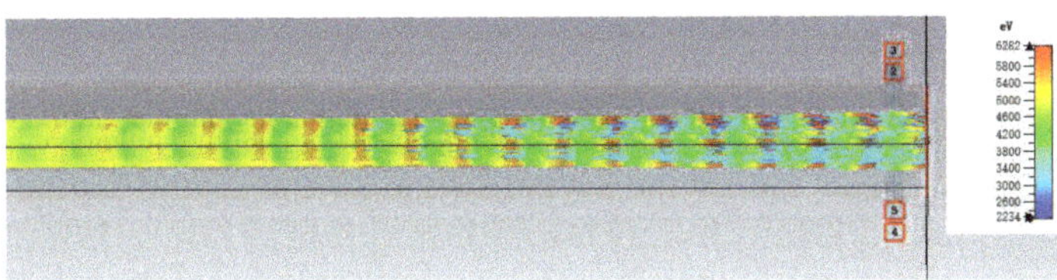

Figure 5. Typical electron bunching along a longitudinal direction.

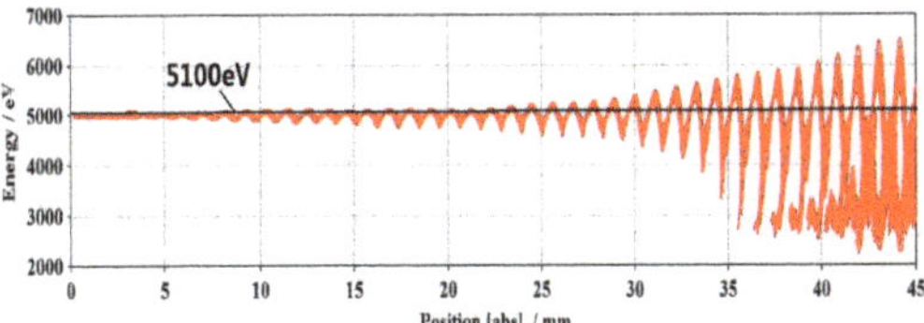

Figure 6. Electron kinetic energy versus longitudinal distance.

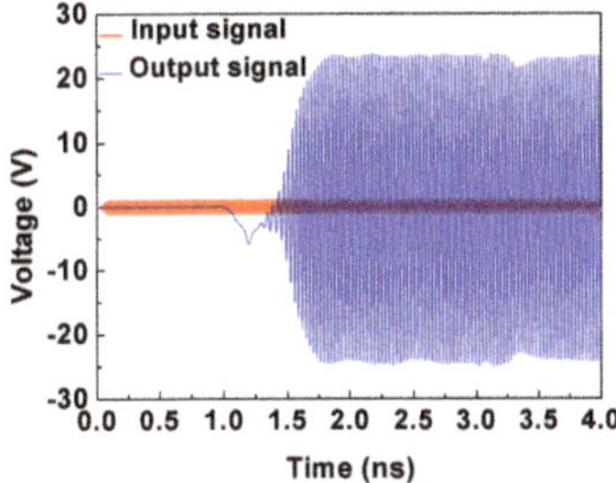

Figure 7. Input and output signals plot of the novel amplifier at 35 GHz.

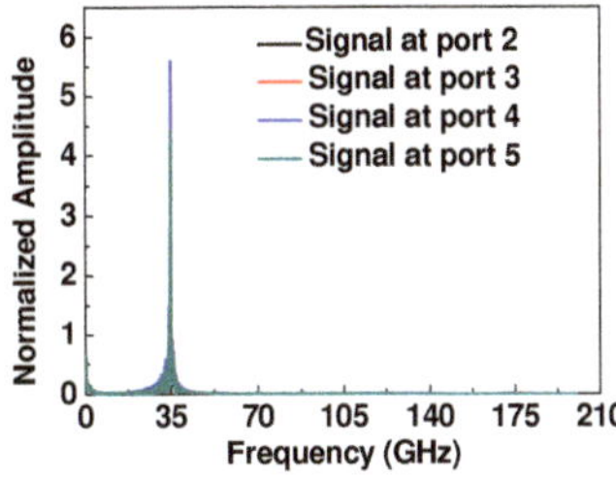

Figure 8. Spectrums of the signals plot of the four output ports.

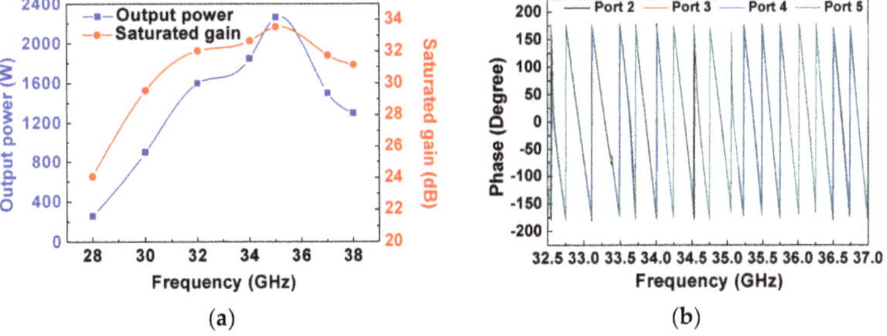

Figure 9. Plots of the simulated output power: the saturated gain (**a**); the phase congruency (**b**) for four output ports versus frequency.

Figure 6 shows the electron kinetic energy versus longitudinal distance when the electron dynamic system has been in steady state. As most of the electrons are decelerated at the end of the circuit, most kinetic energy from the sheet beam is transferred to electromagnetic fields and the excitation signal is amplified.

Figure 7 shows the signal amplitudes of the input port and one of the four output ports. The output signals become stable after a transient time about 1.6 ns. With the energy transfer from the electron beam to the high-frequency field, the input signal with the amplitude of 1 V is amplified to 23.8 V for a single amplifier channel, with the gains more than 27.5 dB. The signal amplitude differences of the four output ports are less than 5% according to our simulations. Therefore, the total gains for the whole 2 × 2 amplifier unit are around 33.5 dB.

As can be seen in Figure 8, the output signal spectrums are concentrated at around 35 GHz and relatively pure. The normalized amplitudes at the frequency points of 35 GHz, 70 GHz, and 105 GHz are 5.62, 0.04, and 0.0006, respectively. Due to the optimization of attenuators, no obvious backward-wave oscillation from higher order modes is observed.

The sum of the average output power of the four output ports and the saturated gain versus frequency of the amplifier according to the simulation results at a frequency of 35 GHz are 2260 W and 33.5 dB, respectively, as shown in Figure 9a. The instantaneous 3-dB bandwidth of the 2 × 2 amplifier units is 7 GHz, which demonstrates that it can maintain the broadband advantage of the MML SWS TWT.

The phase congruency for the four output port signals is investigated for its PAR applications. From the phase analysis results in CST shown in Figure 9b, we can see the phase differences among the four channels are below 5%. It indicates the great phase congruency for different channels.

4. Fabrication and Experimental Test

The "cold" transmission model with 80-period SWS is fabricated and tested to verify the transmission performances preliminarily. Rogers RT5880 is employed as the material of the dielectric substrates. The cavity is designed as a two-halves structure, in which the SWSs are located on the upper and lower slots. As shown in Figure 10a,b, an input waveguide port is on one side of the cavity and four coaxial cable output ports are located on another side. The dimensions of the amplifier are shown in Figure 10c.

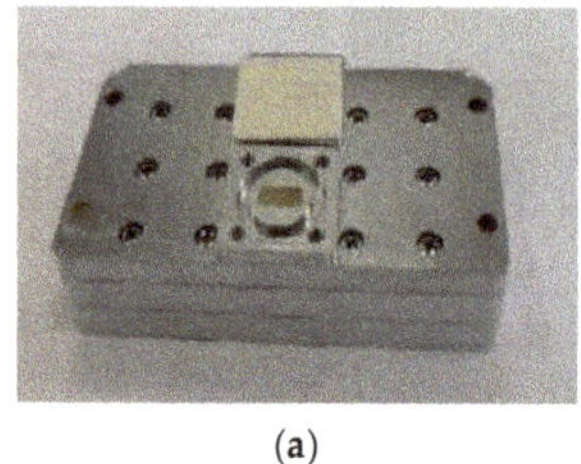

(a)

(b)

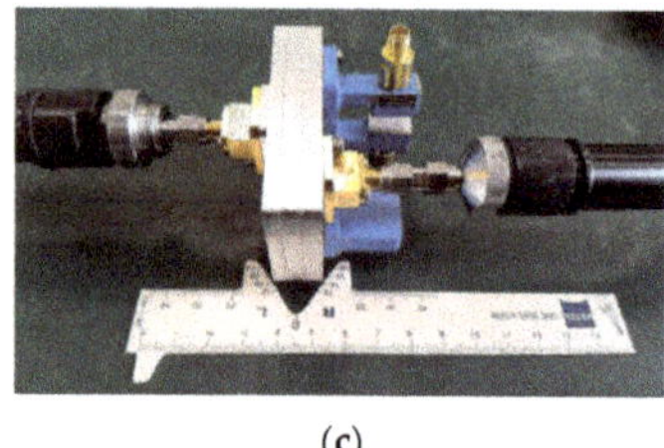

(c)

Figure 10. Fabricated cavity for the transmission model with the input port at the front (**a**); the output ports at the back (**b**); the dimensions (**c**).

Without being driven by the electron beam and input signals, only the "cold" transmission performances of the fabricated 80-period SWS in the cavity are experimentally tested preliminarily by the vector network analyzer. The S-parameters are tested when all other ports are connected with matched loads. The comparisons of the simulated values and the tested values are shown in Figure 11a,b. As shown in Figure 11a, the tested S_{11} values are almost below −15 dB, which is basically consistent with the simulation results. It can be seen in Figure 11b that the tested "cold" transmission loss values are around 7.5 dB, which are 0.5 dB more than the simulation values. The larger loss may be due to assembly errors.

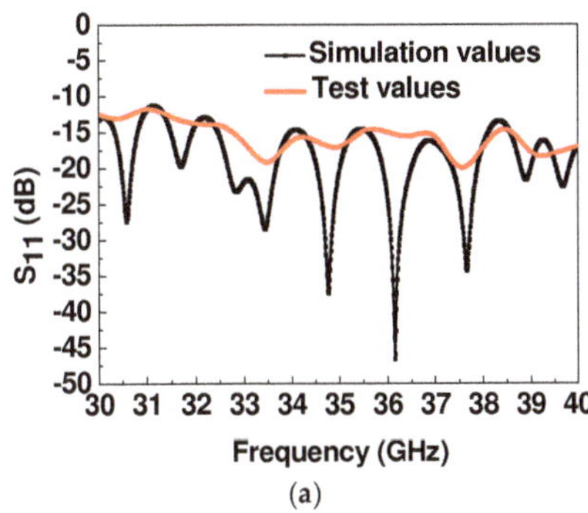

(a)

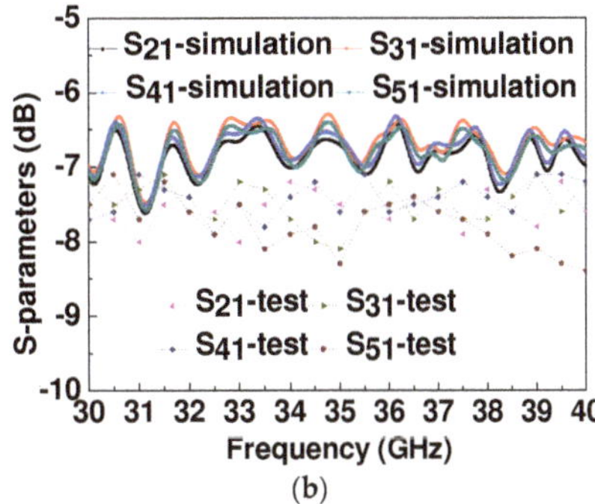

(b)

Figure 11. Comparison between simulation values and test values for S_{11} parameters (**a**); transmission losses (**b**).

5. Conclusions and Discussion

This novel amplifier can probably break the traditional "one TWT for one channel" framework in PAR application. Driven by only one sheet electron beam and one input signal, four independent output signals are generated with more than 2000 W saturated peak power and 7 GHz bandwidth according to the simulation results. The phase congruency among the four output channels is also great. Moreover, this compact framework is easy to integrate with solid-state circuits. The advantages can be summarized as high power, broad bandwidth, good integrability, compact structure, and great phase congruency. Based on the analysis above, it can be concluded that this novel 2 × 2 TWT amplifier unit is promising on the millimeter wave T/R modules of the PARs. Finally, the transmission model is fabricated, and the "cold" transmission performances are tested to show the practicability of this novel amplifier.

Compared with the Ka-band GaAs power amplifier shown in reference [5], the amplifier can supply higher peak output power (2000 W vs. 4 W), higher peak gain (33.5 dB vs. 22 dB) and wider 3-dB bandwidth (7 GHz vs. 5 GHz) according to the simulation results. Compared with the MML TWT power amplifier shown in reference [11–13], the amplifier can supply multi-channel output signals and favorable phase congruency between channels, and thus is more suitable for the PAR applications.

However, there are still some practical issues that need to be discussed. Firstly, the preliminary tests mainly focus on the "cold" S-parameters without an electron beam. The electron gun and focus magnetic field will be considered in the next experimental step. Secondly, as a novel type of vacuum electronic device, the limited range of the working environment is still vacuum condition, so the vacuum treatment needs further research. Thirdly, due to the large power delivery, the thermal dissipation structures should be designed specially in applications.

Author Contributions: Conceptualization, G.G. and Y.W.; methodology, J.L.; software, Z.S.; validation, Z.Y. and Z.S.; writing—original draft preparation, R.Y.; writing—review and editing, G.G.; visualization, R.Y.; supervision, Y.G.; project administration, Y.W.; funding acquisition, G.G. All authors have read and agreed to the published version of the manuscript.

Funding: This work was supported in part by the Fundamental Research Funds for the Central Universities under Grant No. ZYGX2018J032, in part by the National Natural Science Foundation of China (Grant No. 61801088, 92163204, 61921002) and in part by the National Key R&D Program of China, contract numbers 2017YFE0300200 and 2017YFE0300201.

Data Availability Statement: The data presented in this study are available on request from the corresponding author.

Conflicts of Interest: The authors declare no conflict of interest.

References

1. Bil, R.; Holpp, W. Modern Phased Array Radar Systems in Germany. In Proceedings of the 2016 IEEE International Symposium on Phased Array Systems and Technology(PAST), Waltham, MA, USA, 18–21 October 2016; pp. 1–7.
2. Herd, J.; Carlson, D.; Duffy, S.; Weber, M.; Brigham, G.; Rachlin, M.; Cursio, D.; Liss, C.; Weigand, C. Multifunction Phased Array Radar (MPAR) for Aircraft and Weather Surveillance. In Proceedings of the 2010 IEEE Radar Conference, Arlington, VA, USA, 10–14 May 2010; pp. 945–948.
3. Turlington, T.R.; Sacks, F.E.; Gipprich, J.W. T/R Module Architectural Consideration for Active Electronically Steerable Arrays. In Proceedings of the 1992 IEEE MTT-S Microwave Symposium Digest, Albuquerque, NM, USA, 1–5 June 1992; Volume 3, pp. 1523–1526.
4. Turlington, T.R. Additive Functions Provide a Powerful Tool for T/R Module Modeling. In Proceedings of the 2007 IEEE Radar Conference, Boston, MA, USA, 17–20 April 2007; pp. 954–959.
5. Colomb, F.Y.; Platzker, A. 2 and 4 watt Ka-band GaAs PHEMT power amplifier MMICs. In Proceedings of the IEEE MTT-S International Microwave Symposium Digest, Philadelphia, PA, USA, 8–13 June 2003; Volume 2, pp. 843–846.
6. Pelk, M.J.; Neo, W.C.E.; Gajadharsing, J.R.; Pengelly, R.S.; de Vreede, L.C.N. A High-Efficiency 100-W GaN Three-Way Doherty Amplifier for Base-Station Applications. *IEEE Trans. Microw. Theory Tech.* **2008**, *56*, 1582–1591. [CrossRef]
7. Booske, J.H.; Dobbs, R.J.; Joye, C.D.; Kory, C.L.; Neil, G.R.; Park, G.-S.; Park, J.; Temkin, R.J. Vacuum Electronic High Power Terahertz Sources. *IEEE Trans. Terahertz Sci. Technol.* **2011**, *1*, 54–75. [CrossRef]
8. Baig, A.; Shin, Y.; Barnett, L.R.; Gamzina, D.; Barchfeld, R.; Domier, C.W.; Wang, J.; Luhmann, N.C., Jr.; Ieee, F. Design, Fabrication and RF Testing of Near-THz Sheet Beam TWTA. *IEEE Trans. Terahertz Sci. Technol.* **2011**, *4*, 181–207.
9. Gamzina, D.; Himes, L.G.; Barchfeld, R.; Zheng, Y.; Popovic, B.K.; Paoloni, C.; Choi, E.; Luhmann, N.C. Nano-CNC Machining of Sub-THz Vacuum Electron Devices. *IEEE Trans. Electron Devices* **2016**, *63*, 4067–4073. [CrossRef]
10. Shen, F.; Wei, Y.; Yin, H.; Gong, Y.; Xu, X.; Wang, S.; Wang, W.; Feng, J. A Novel V-Shaped Microstrip Meanderline Slow-Wave Structure For W-Band MMPM. *IEEE Trans. Plasma Sci.* **2012**, *40*, 463–469. [CrossRef]
11. Sengele, S.; Jiang, H.; Booske, J.H.; van der Weide, D.; Kory, C.; Ives, L. A Selectively Metallized, Microfabricated W-Band Meanderline TWT Circuit. In Proceedings of the 2008 IEEE International Vacuum Electronics Conference, Monterey, CA, USA, 22–24 April 2008; pp. 447–448.
12. Shen, F.; Wei, Y.Y.; Xu, X.; Liu, Y.; Huang, M.Z.; Tang, T.; Duan, Z.Y.; Gong, Y.B. Symmetric Double V-Shaped MML Slow-Wave Structure for W-Band Traveling-Wave Tube. *IEEE Trans. Electron Devices* **2012**, *59*, 1551–1557. [CrossRef]
13. Nehra, A.K.; Gupta, R.K.; Panda, P.C.; Sharma, S.M.; Choyal, Y.; Sharma, R.K. Electron Gun Design for Multi-Beam Pulsed Amplifier. In Proceedings of the IEEE International Vacuum Electronics Conference, Monterey, CA, USA, 24–26 April 2012; pp. 175–176.
14. Dassault Systemes. CST-Computer Simulation Technology. 2020. Available online: http://www.cst.com (accessed on 25 September 2021).

 MDPI

Article

Green's Functions of Multi-Layered Plane Media with Arbitrary Boundary Conditions and Its Application on the Analysis of the Meander Line Slow-Wave Structure

Zheng Wen [1,2], Jirun Luo [1,2,*] and Wenqi Li [1]

[1] Key Laboratory of Science and Technology on High Power Microwave Sources and Technologies, Aerospace Information Research Institute, Chinese Academy of Sciences, Beijing 100190, China; wenzheng17@mails.ucas.edu.cn (Z.W.); liwenqi16@mails.ucas.ac.cn (W.L.)

[2] School of Electronic, Electrical and Communication Engineering, University of Chinese Academy of Sciences, Beijing 100039, China

* Correspondence: luojirun@mail.ie.ac.cn

Abstract: A method was proposed for solving the dyadic Green's functions (DGF) and scalar Green's functions (SGF) of multi-layered plane media in this paper. The DGF and SGF were expressed in matrix form, where the variables of the boundary conditions (BCs) can be separated in matrix form. The obtained DGF and SGF are in explicit form and suitable for arbitrary boundary conditions, owing to the matrix form expression and the separable variables of the BCs. The Green's functions with typical BCs were obtained, and the dispersion characteristic of the meander line slow-wave structure (ML-SWS) is analyzed based on the proposed DGF. The relative error between the theoretical results and the simulated ones with different relative permittivity is under 3%, which demonstrates that the proposed DGF is suitable for electromagnetic analysis to complicated structure including the ML-SWS.

Keywords: dyadic Green's functions; inhomogeneous; multi-layered media; slow-wave structures

Citation: Wen, Z.; Luo, J.; Li, W. Green's Functions of Multi-Layered Plane Media with Arbitrary Boundary Conditions and Its Application on the Analysis of the Meander Line Slow-Wave Structure. *Electronics* **2021**, *10*, 2716. https://doi.org/10.3390/electronics10212716

Academic Editor: Giovanni Andrea Casula

Received: 14 October 2021
Accepted: 4 November 2021
Published: 8 November 2021

1. Introduction

Since the exact results of the two-layered planar dielectric model were deduced [1], more and more people have been engaged in research of the electromagnetic field for multi-layered media [2–22], which has been widely utilized for the analysis of dielectric waveguides, printed circuit boards, antennas and sensors [23–26].

The Green's functions, including dyadic Green's functions (DGF) and scalar Green's functions (SGF), are powerful tools in electromagnetic theory [2–4], because the relationship between the field and excitation sources can be easily described by them. As a result, many years of effort have been devoted to obtaining Green's functions for inhomogeneous media [3–22] There are many approximation and numerical methods that have been utilized for calculating the DGF and SGF, such as the finite sum superposition method [3], the total least squares method [4], the fast full-mode method [5], the numerical modified steepest descent path method [5] and the numerically stable analysis method [6]. In addition, the pure theoretical derivations of Green's functions for the stratified media, without numerical approximation and error, were also pursued for a long time [7–22]

The form of the dyadic Green's function has been presented based on the methods of generation function expansions [7–15] (the generation functions were called vector wave functions in [7–9], eigen-functions in [10,12–15] and solenoidal Hausen vectors in [11]), and this form can also be used in multi-layered media conditions [8–15]. If these traditional generation functions, such as those in [8], are selected to obtain the DGF, the boundary conditions (BCs) between the adjacent layers will be satisfied by another method, such as the method of scattering superposition [9,10]. Consequently, the equations built from the boundary conditions may be complicated and the results may be not in explicit form.

The method of scattering superposition is direct with explicit physical meanings for multi-layered media [7]. However, it will be complicated when the number of layers is increased. Instead, if the BCs has been contained in the generation function with explicit form, the corresponding Green's function can be simple and in explicit form.

In addition, methods such as the perturbative approach [16], operator theories [18–20], wave superposition [21] and transmission line theories [17,18,21,22] are also introduced for obtaining the Green's function and exactly analyzing the multi-layered media. Unfortunately, the applications of these results are limited, owing to the specified boundary conditions at two ends of the multi-layered media [18]. In [21], although the DGF of multi-layered media is obtained with undefined boundary conditions at two ends, the DGF is deduced in the rectangular waveguide, whose length is assumed infinite, and the direction of the stratified media must be defined in consistence with that of the guided wave's propagation. In [20–22], the obtained Green's function of multi-layered plane media contain the BCs between the adjacent layers, but they do not contain the BCs at the top and the bottom of multi-layered media. When they are utilized to analyze the specified structures, the BCs at two ends of multi-layered media must be considered, and the corresponding equations will be built from the BCs and these Green's functions [18]. As a result, the equations of the structure base on the Green's function in [18] may be more complicated, because BCs at two ends may add the number of the equation.

In this paper, a method was proposed for obtaining the Green's functions of multi-layered media with arbitrary boundary conditions, including dyadic Green's functions and scalar ones. In this method, the Green's functions consist of the generation functions. In the process of deducing the generation functions, the BCs between the adjacent layers and at two ends have been considered, and are represented by a series of variables. These variables of the BCs can be separated in matrix form. Correspondingly, the DGF and SGF were obtained in explicit form with matrixes, and corresponding equations built from BCs at the top and the bottom of the multi-layered media are independent and easy to solve, which means that the equations of the structure based on the proposed DGF may be more simple with clearer physical characteristics and that the formulae may be expressed with the computer code more friendly.

In Section 2, the proposed method was discussed. The typical BCs, such as metal boundary conditions, radiation boundary conditions and their combinations, were discussed in Section 3, respectively. Furthermore, the application on the dispersion analysis of a meander line slow-wave structure is given in Section 4. The effect of relative permittivity on the dispersion is discussed, and the results of theoretical calculation are compared with those of the simulation. In addition, conclusions are drawn in Section 5.

It should be noted that, in this paper, the y-axis is regarded as the referent direction, k_0 is the wave number in free space, k_c is the eigenvalue in the cross section to referent direction, and $\delta_{i,j}$ is the Kronecker delta function. Moreover, the symbol $\sum$ will be replaced by $\int$, if the corresponding eigenvalue is continuous.

2. The DGF and SGF of Multi-Layered Plane Media with Arbitrary Boundaries

Figure 1 shows the geometry of multi-layered media, which is stratified along the y-axis and can be divided into N layers.

$$\varepsilon_r = \varepsilon_{r,j} \text{ when } y \in (y_{j-1}, y_j] j = 1, 2, \ldots, N \tag{1}$$

where ε_r ($\varepsilon_{r,j}$) is the relative permittivity (in the jth layer). y_{j-1} and y_j are the two edge values of the jth layer at y-axis. The two bottom boundaries are marked "Boundary 1" and "Boundary 2", respectively.

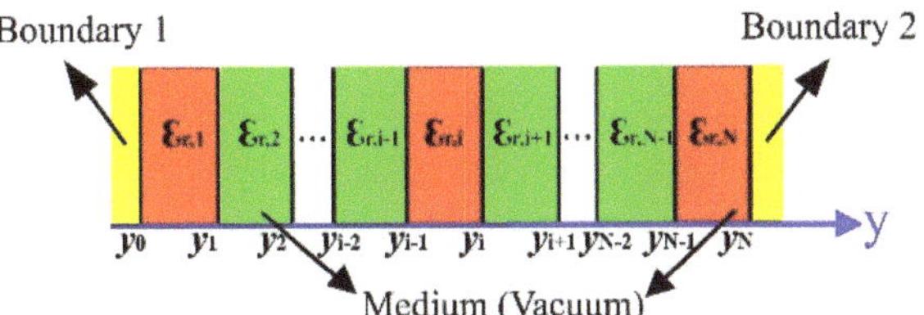

Figure 1. Geometry of multi-layered media.

Owing to the principle of field superposition, most sources can be regarded as a sum of point sources. Therefore, a 3D point source δ is selected as the excitation source. Assuming that the excitation source, marked with "′", is in the ith layer, and choosing y-axis as the referent direction, the SGF $G^j\left(\vec{R},\vec{R}'\right)$ for the jth layer satisfies the Helmholtz equation as follows:

$$\begin{cases} \nabla^2 G^j\left(\vec{R},\vec{R}'\right) + \varepsilon_{r,j}k_0^2 G^j\left(\vec{R},\vec{R}'\right) = 0, j \neq i, j = 1,2,\ldots,N \\ \nabla^2 G^i\left(\vec{R},\vec{R}'\right) + \varepsilon_{r,i}k_0^2 G^i\left(\vec{R},\vec{R}'\right) = -\delta\left(\vec{R}-\vec{R}'\right), j = i \end{cases} \quad (2)$$

Furthermore, the GF can be expressed as the sum of TM and TE components to the referent direction (y-axis), namely,

$$G^j\left(\vec{R},\vec{R}'\right) = G^j_m\left(\vec{R},\vec{R}'\right) + G^j_n\left(\vec{R},\vec{R}'\right) \quad (3)$$

The boundary conditions at $y = y_j$ can be written as:

$$\begin{cases} \sqrt{\varepsilon_{r,j}}G^j_m\left(\vec{R},\vec{R}'\right) = \sqrt{\varepsilon_{r,j+1}}G^{j+1}_m\left(\vec{R},\vec{R}'\right), j = 1,2,\ldots,N-1 \\ \frac{1}{\sqrt{\varepsilon_{r,j}}}\frac{\partial}{\partial y}G^j_m\left(\vec{R},\vec{R}'\right) = \frac{1}{\sqrt{\varepsilon_{r,j+1}}}\frac{\partial}{\partial y}G^{j+1}_m\left(\vec{R},\vec{R}'\right) \\ G^j_n\left(\vec{R},\vec{R}'\right) = G^{j+1}_n\left(\vec{R},\vec{R}'\right) \\ \frac{\partial}{\partial y}G^j_n\left(\vec{R},\vec{R}'\right) = \frac{\partial}{\partial y}G^{j+1}_n\left(\vec{R},\vec{R}'\right) \end{cases} \quad (4)$$

Because the shape of cross section is uniform along the y-axis, considering the isotropic and lossless media, $G^j_{\substack{m\\n}}\left(\vec{R},\vec{R}'\right)$ can be expressed according to function $f^i_{\substack{m\\n},j}(y,y')$ for different transverse eigenvalues k_c [18].

$$G^j_{\substack{m\\n}}\left(\vec{R},\vec{R}'\right) = \sum_{k_c} g_{\substack{m\\n}} \cdot f^i_{\substack{m\\n},j}(y,y') \quad (5)$$

where, index i represents that the excitation source is in the ith layer and $g_{\substack{m\\n}}$ is the coefficient of the series.

The $f^i_{\substack{m\\n},j}(y,y')$ in Equation (5) can be described as follows:

$$\frac{d^2}{dy^2}f^i_{\substack{m\\n},j}(y,y') + \left(\varepsilon_{r,j}k_0^2 - k_c^2\right)f^i_{\substack{m\\n},j}(y,y') = 0 \text{ when } y \in (y_{j-1},y_j), j \neq i, j = 1,2,\ldots,N \quad (6)$$

$$\frac{d^2}{dy^2}f^i_{\substack{m\\n},i}(y,y') + \left(\varepsilon_{r,i}k_0^2 - k_c^2\right)f^i_{\substack{m\\n},i}(y,y') = -F_{\substack{m\\n}}g^*_{\substack{m\\n}}(x',z')\delta(y-y') \text{ when } y \in (y_{i-1},y_i) \quad (7)$$

where the $F_{\substack{m\\n}}$ is the normalized coefficient, and Equation (4) can be simplified, when $y = y_j$

$$\begin{cases} \sqrt{\varepsilon_{r,j}} f^i_{m,j}(y,y') = \sqrt{\varepsilon_{r,j+1}} f^i_{m,j+1}(y,y'), f^i_{n,j}(y,y') = f^i_{n,j+1}(y,y') \\ \frac{1}{\sqrt{\varepsilon_{r,j}}} \frac{d}{dy} f^i_{m,j}(y,y') = \frac{1}{\sqrt{\varepsilon_{r,j+1}}} \frac{d}{dy} f^i_{m,j+1}(y,y'), \frac{d}{dy} f^i_{n,j}(y,y') = \frac{d}{dy} f^i_{n,j+1}(y,y') \end{cases} \tag{8}$$

According to the operator theories, the electromagnetic field can be analyzed in the Hilbert space, and the function space consisting of orthonormal basis functions, $\{ \sin(ky) \quad \cos(ky) \}$, is complete [18].

Assuming that the $f^i_{\substack{m\\n},j}(y,y')$ can be written as:

$$\begin{cases} f^i_{\substack{m\\n},k}(y,y') = C_{\substack{m\\n}} \left[S^+_{y,k}(y)\right] \left[\begin{array}{cc} a_{\substack{m\\n},k} & b_{\substack{m\\n},k} \end{array} \right]^T, y_{k-1} \le y \le y_k, y \le y' \\ f^i_{\substack{m\\n},l}(y,y') = C_{\substack{m\\n}} \left[S^-_{y,l}(y)\right] \left[\begin{array}{cc} c_{\substack{m\\n},l} & d_{\substack{m\\n},l} \end{array} \right]^T, y_{l-1} \le y \le y_l, y' \le y \end{cases} \tag{9}$$

where, the $C_{\substack{m\\n}}$ is the weight factor, $1 \le k \le i \le l \le N$,

$$C_{\substack{m\\n}} = F_{\substack{m\\n}} g^*_{\substack{m\\n}}(x',z') \tag{10}$$

$$k_{y,i} = \left(\varepsilon_{r,i} k_0^2 - k_c^2\right) \tag{11}$$

$$\begin{cases} \left[S^+_{y,j}(y)\right] = \left[\begin{array}{cc} \sin(k_{y,j}(y - y_{j-1})) & \cos(k_{y,j}(y - y_{j-1})) \end{array} \right] \\ \left[S^-_{y,j}(y)\right] = \left[\begin{array}{cc} \sin(k_{y,j}(y_j - y)) & \cos(k_{y,j}(y_j - y)) \end{array} \right] \end{cases} \tag{12}$$

Putting Equations (9)–(12) into Equation (8), one can get:

$$\left[\begin{array}{c} a_{\substack{m\\n},k+1} \\ b_{\substack{m\\n},k+1} \end{array} \right] = \left[T_{\substack{m\\n},k}\right] \left[\begin{array}{c} a_{\substack{m\\n},k} \\ b_{\substack{m\\n},k} \end{array} \right] \quad , \quad \left[\begin{array}{c} c_{\substack{m\\n},l} \\ d_{\substack{m\\n},l} \end{array} \right] = \left[R_{\substack{m\\n},l}\right] \left[\begin{array}{c} c_{\substack{m\\n},l+1} \\ d_{\substack{m\\n},l+1} \end{array} \right] \tag{13}$$

where matrixes $[T_{m,k}]$, $[T_{n,k}]$, $[R_{m,l}]$ and $[R_{n,l}]$ are provided in Appendix A.

For brief expression, matrixes $\left[A_{\substack{m\\n},i}\right]$ and $\left[B_{\substack{m\\n},i}\right]$ can be defined:

$$\left[A_{\substack{m\\n},i}\right] = T_{\substack{m\\n},i-1} T_{\substack{m\\n},i-2} \cdots T_{\substack{m\\n},1} T_{\substack{m\\n},0} = \left[\begin{array}{cc} A_{\substack{m\\n},i1} & A_{\substack{m\\n},i2} \\ A_{\substack{m\\n},i3} & A_{\substack{m\\n},i4} \end{array} \right], \left[B_{\substack{m\\n},i}\right] = R_{\substack{m\\n},i} R_{\substack{m\\n},i+1} \cdots R_{\substack{m\\n},N-1} R_{\substack{m\\n},N} = \left[\begin{array}{cc} B_{\substack{m\\n},i1} & B_{\substack{m\\n},i2} \\ B_{\substack{m\\n},i3} & B_{\substack{m\\n},i4} \end{array} \right] \tag{14}$$

Therefore, Equation (9) can be rewritten as

$$f^i_{\substack{m\\n},j}(y,y') = C_{\substack{m\\n}} \begin{cases} \left[S^+_{y,j}(y)\right] \left[A_{\substack{m\\n},j}\right] \left[\begin{array}{cc} a_{\substack{m\\n},1} & b_{\substack{m\\n},1} \end{array} \right]^T, y_{j-1} \le y \le y_j, y \le y' \\ \left[S^-_{y,j}(y)\right] \left[B_{\substack{m\\n},j}\right] \left[\begin{array}{cc} c_{\substack{m\\n},N} & d_{\substack{m\\n},N} \end{array} \right]^T, y_{j-1} \le y \le y_j, y' \le y \end{cases} \tag{15}$$

Here, the $f^i_{\substack{m\\n},j}(y,y')$ can be represented as:

$$f^i_{m,j}(y,y') = C_m M_j(y) \left(U^*_i(y')\right)^T, f^i_{n,j}(y,y') = C_n N_j(y) \left(V^*_i(y')\right)^T \tag{16}$$

where,

$$\begin{cases} M_j(y) = \begin{cases} \left[S^+_{y,j}(y)\right]\left[A_{m,j}\right], y \le y' \\ \left[S^-_{y,j}(y)\right]\left[B_{m,j}\right], y \ge y' \end{cases}, y_{j-1} \le y \le y_j \\ U_i^*(y') = \begin{cases} \left[\begin{array}{cc} a_{m,1} & b_{m,1} \end{array}\right], y \le y' \\ \left[\begin{array}{cc} c_{m,N} & d_{m,N} \end{array}\right], y \ge y' \end{cases}, y_{i-1} \le y' \le y_i \\ N_j(y) = \begin{cases} \left[S^+_{y,j}(y)\right]\left[A_{n,j}\right], y \le y' \\ \left[S^-_{y,j}(y)\right]\left[B_{n,j}\right], y \ge y' \end{cases}, y_{j-1} \le y \le y_j \\ V_i^*(y') = \begin{cases} \left[\begin{array}{cc} a_{n,1} & b_{n,1} \end{array}\right], y \le y' \\ \left[\begin{array}{cc} c_{n,N} & d_{n,N} \end{array}\right], y \ge y' \end{cases}, y_{i-1} \le y' \le y_i \end{cases} \tag{17}$$

As a result, SGF $G^j\left(\vec{R}, \vec{R}'\right)$ can be represented as

$$G^j\left(\vec{R}, \vec{R'}\right) = \sum_{k_c}\left(g_m(x,z)\cdot\left(C_m M_j(y)\left(U_i^*(y')\right)^T\right) + g_n(x,z)\cdot\left(C_n N_j(y)\left(V_i^*(y')\right)^T\right)\right) \tag{18}$$

Furthermore, consider the Equation (10) and define the generating functions as:

$$\begin{cases} \varphi_m\left(\vec{R}\right) = g_m(x,z)M_j(y), \phi^*_m\left(\vec{R'}\right) = g^*_m(x',z')U_i^*(y') \\ \varphi_n\left(\vec{R}\right) = g_n(x,z)N_j(y), \phi^*_n\left(\vec{R'}\right) = g^*_n(x',z')V_i^*(y') \end{cases} \tag{19}$$

where $g^*_m(x,z)$ and $g^*_n(x,z)$ are the conjugate functions of $g_m(x,z)$ and $g_n(x,z)$, respectively. Note that the concrete mathematical form of $g_{\substack{m\\n}}(x,z)$ depends on coordinate systems and boundary conditions in the x-z plane. In a Cartesian coordinate system, $g_{\substack{m\\n}}(x,z)$ can be expressed by basic functions systems $\{\ \sin(kx) \quad \cos(kx)\ \}$ or $\left\{e^{ikx}\right\}$. Then, the SGF $G^j\left(\vec{R}, \vec{R}'\right)$ can be written in a usual expression:

$$G^j\left(\vec{R}, \vec{R'}\right) = \sum_{k_c}\left(F_m\varphi_m\left(\vec{R}\right)\left(\phi^*_m\left(\vec{R'}\right)\right)^T + F_n\varphi_n\left(\vec{R}\right)\left(\phi^*_n\left(\vec{R'}\right)\right)^T\right) \tag{20}$$

In addition, the DGF $\vec{\vec{G}}^j\left(\vec{R}, \vec{R}'\right)$ can be expressed in the same form as that in [9]:

$$\begin{aligned} \vec{\vec{G}}^j\left(\vec{R}, \vec{R'}\right) &= -\frac{\hat{y}\hat{y}}{\varepsilon_{r,i}k_0^2}\delta\left(\vec{R}-\vec{R'}\right) + \sum_{k_c}\left(\frac{F_n}{k_c^2}\left(\left(\nabla\times\left(\varphi_n\left(\vec{R}\right)\hat{y}\right)\right)\left(\nabla'\times\left(\left(\phi^*_n\left(\vec{R'}\right)\right)^T\hat{y}\right)\right)\right)\right) \\ &+\sum_{k_c}\left(\frac{F_m}{k_c^2k_0^2\sqrt{\varepsilon_{r,j}}\sqrt{\varepsilon_{r,j}}}\left(\nabla\times\nabla\times\left(\varphi_m\left(\vec{R}\right)\hat{y}\right)\right)\left(\nabla'\times\nabla'\times\left(\left(\phi^*_m\left(\vec{R'}\right)\right)^T\hat{y}\right)\right)\right) \end{aligned} \tag{21}$$

Considering normalized conditions, the point source equations could be written as below:

$$\begin{cases} f^i_{\substack{m\\n},k}(y,y') - f^i_{\substack{m\\n},l}(y,y') = 0 \\ \frac{d}{dy}f^i_{\substack{m\\n},k}(y,y') - \frac{d}{dy}f^i_{\substack{m\\n},l}(y,y') = 1 \end{cases}, y = y', k = l = i \tag{22}$$

then

$$\left[S_{y,i}^{+}(y')\right]^{T}\left[A_{\substack{m\\n},i}\right]\begin{bmatrix} a_{\substack{m\\n},1} \\ b_{\substack{m\\n},1} \end{bmatrix} = \left[S_{y,i}^{-}(y')\right]^{T}\left[B_{\substack{m\\n},i}\right]\begin{bmatrix} c_{\substack{m\\n},N} \\ d_{\substack{m\\n},N} \end{bmatrix} = k_{y,i}\begin{bmatrix} c_{\substack{m\\n},N} & d_{\substack{m\\n},N} \end{bmatrix}\left[B_{\substack{m\\n},i}\right]^{T}[D_i]\left[A_{\substack{m\\n},i}\right]\begin{bmatrix} a_{\substack{m\\n},1} & b_{\substack{m\\n},1} \end{bmatrix}^{T} \tag{23}$$

where $[D_i]$ is given in Appendix A.

Assume that variables $b_{m,1}$, $d_{m,N}$, $a_{n,1}$, and $c_{n,N}$ are not zero, and they can be presented easily in matrix form:

$$\begin{cases} b_{m,1} = \left[S_{y,i}^{-}(y')\right][B_{m,i}]\begin{bmatrix} \frac{c_{m,N}}{d_{m,N}} & 1 \end{bmatrix}^{T} / (k_{y,i}I_{m,i}), d_{m,N} = \left[S_{y,i}^{+}(y')\right][A_{m,i}]\begin{bmatrix} \frac{a_{m,1}}{b_{m,1}} & 1 \end{bmatrix}^{T} / (k_{y,i}I_{my,i}) \\ a_{n,1} = \left[S_{y,i}^{-}(y')\right][B_{n,i}]\begin{bmatrix} 1 & \frac{d_{n,N}}{c_{n,N}} \end{bmatrix}^{T} / (k_{y,i}I_{n,i}), c_{n,N} = \left[S_{y,i}^{+}(y')\right][A_{n,i}]\begin{bmatrix} 1 & \frac{a_{n,1}}{b_{n,1}} \end{bmatrix}^{T} / (k_{y,i}I_{ny,i}) \end{cases} \tag{24}$$

$$I_{my,i} = \begin{bmatrix} \frac{c_{m,N}}{d_{m,N}} & 1 \end{bmatrix}[B_{m,i}]^{T}[D_i][A_{m,i}]\begin{bmatrix} \frac{a_{m,1}}{b_{m,1}} & 1 \end{bmatrix}^{T}, I_{ny,i} = \begin{bmatrix} 1 & \frac{d_{n,N}}{c_{n,N}} \end{bmatrix}[B_{n,i}]^{T}[D_i][A_{n,i}]\begin{bmatrix} 1 & \frac{b_{n,1}}{a_{n,1}} \end{bmatrix}^{T} \tag{25}$$

It is worth noting that variables $a_{m,1}$, $b_{m,1}$, $c_{m,N}$, $d_{m,N}$, $a_{n,1}$, $b_{n,1}$, $c_{n,N}$ and $d_{n,N}$ can be exactly deduced by the source conditions Equation (23) and conditions of boundary 1 and 2.

3. Typical Boundary Conditions and Examples

3.1. Metal Boundary Conditions

The typical boundaries are metal conditions. If boundary 1 and 2 are regarded as the metal conditions, the boundary equations can be derived as

$$\frac{d}{dy}f_{m,j}^{i}(y,y') = 0, \text{ and } f_{n,j}^{i}(y,y') = 0, j = 1, y = y_0 \text{ } or \text{ } j = N, y = y_N \tag{26}$$

As a consequence,

$$a_{m,1} = c_{m,N} = b_{n,1} = d_{n,N} = 0 \tag{27}$$

Bringing Equation (27) into source Equations (24) and (25), the rest of variables can be confirmed:

$$\begin{cases} b_{m,1} = \left[S_{y,i}^{-}(y')\right][B_{m,i}]\begin{bmatrix} 0 & 1 \end{bmatrix}^{T} / (k_{y,i}I_{m,i}), d_{m,N} = \left[S_{y,i}^{+}(y')\right][A_{m,i}]\begin{bmatrix} 0 & 1 \end{bmatrix}^{T} / (k_{y,i}I_{m,i}) \\ a_{n,1} = \left[S_{y,i}^{-}(y')\right][B_{n,i}]\begin{bmatrix} 1 & 0 \end{bmatrix}^{T} / (k_{y,i}I_{n,i}), c_{n,N} = \left[S_{y,i}^{+}(y')\right][A_{n,i}]\begin{bmatrix} 1 & 0 \end{bmatrix}^{T} / (k_{y,i}I_{n,i}) \end{cases} \tag{28}$$

where

$$I_{m,i} = \begin{bmatrix} 0 & 1 \end{bmatrix}[B_{m,i}]^{T}[D_i][A_{m,i}]\begin{bmatrix} 0 & 1 \end{bmatrix}^{T}, I_{n,i} = \begin{bmatrix} 1 & 0 \end{bmatrix}[B_{n,i}]^{T}[D_i][A_{n,i}]\begin{bmatrix} 1 & 0 \end{bmatrix}^{T} \tag{29}$$

Then the SGF and the DGF can be written easily, according to Equations (20) and (21).

Example: A Rectangular Waveguide Laterally Filled with Multi-Layered Media

In Figure 2, the rectangular waveguide is laterally filled with multi-layered plane media along the y-axis. The guided wave is along the z-axis. As a result,

$$g_m(x,z) = \sin(k_x x)e^{ihz}, g_n(x,z) = \cos(k_x x)e^{ihz}, F_m = 2/(\pi b), F_n = 2(2 - \delta_{k_c,0})/(\pi b), k_x = p\pi/b \tag{30}$$

where b is the height of the rectangular waveguide at x-axis.

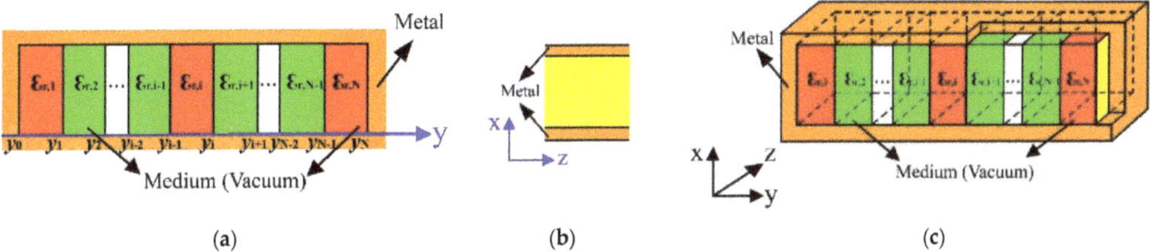

(a) (b) (c)

Figure 2. Geometry of the rectangular waveguide laterally filled with multi-layered media:(**a**) front view; (**b**) side view; (**c**) 3D model view.

As mentioned above, the DGF and GF can be exactly given by Equations (20) and (21). Moreover, one can define N = 2, $k_{y,1} = \beta_2$, $y_0 = 0$, $y_1 = d$, $y_2 = a$, $\varepsilon_{r,2}k_0^2 = k_1^2$, $\varepsilon_{r,1}k_0^2 = k_2^2$, $k_c^2 = k_x^2 + h^2$, $F = 1/(2\pi b)$, and the excitation source in the second layer (i = 2). The DGF in the first layer can be written as:

$$\overset{\rightarrow}{\overset{\rightarrow}{G^1}}\left(\vec{R},\vec{R'}\right) = \sum_P \int \frac{\left(\begin{array}{c}\frac{\left(\nabla\times\left(\sin\left(k_{y,1}(y-y_0)\right)g_n(x,z)\hat{y}\right)\right)\left(\nabla'\times\left(\sin\left(k_{y,2}(y_2-y')\right)g_n^*(x',z')\hat{y}\right)\right)}{\left(k_{y,2}\sin\left(k_{y,1}(y_1-y_0)\right)\cos\left(k_{y,2}(y_2-y_1)\right)+k_{y,1}\cos\left(k_{y,1}(y_1-y_0)\right)\sin\left(k_{y,2}(y_2-y_1)\right)\right)} \\ +\frac{\left(\nabla\times\nabla\times\left(\cos\left(k_{y,1}(y-y_0)\right)g_m(x,z)\hat{y}\right)\right)\left(\nabla'\times\nabla'\times\left(\cos\left(k_{y,2}(y_2-y')\right)g_m^*(x',z')\hat{y}\right)\right)}{\left(\varepsilon_{r,2}k_{y,1}\sin\left(k_{y,1}(y_1-y_0)\right)\cos\left(k_{y,2}(y_2-y_1)\right)+\varepsilon_{r,1}k_{y,2}\cos\left(k_{y,1}(y_1-y_0)\right)\sin\left(k_{y,2}(y_2-y_1)\right)\right)}\end{array}\right)}{\pi b k_0^2((p\pi/b)^2+h^2)(1+\delta_0)} dh \quad (31)$$

which is consistent with the results in [7].

As for the rectangular cavity filled with multi-layered plane media, Equation (30) is supposed to be replaced as follows, respectively:

$$g_m(x,z) = \sin(k_x x)\sin(hz), g_n(x,z) = \cos(k_x x)\cos(hz), F_m = 4/(ab), F_n = 4(2-\delta_{k_c,0})/(ab), k_x = p\pi/b, h = q\pi/a \quad (32)$$

where a is the width of the rectangular cavity at z-axis.

3.2. Infinite Radiation Boundary Conditions

While both boundaries 1 and 2 in Figure 1 are infinite radiation boundaries, the boundary equations are

$$\lim_{y_0\to-\infty}\left(y^{\frac{s-1}{2}}\left(\frac{d}{dy}f^i_{\substack{m\\n},1}(y,y') + ik_{y,1}f^i_{\substack{m\\n},1}(y,y')\right)\right)\Bigg|_{y=y_0} = 0, \lim_{y_N\to\infty}\left(y^{\frac{s-1}{2}}\left(\frac{d}{dy}f^i_{\substack{m\\n},N}(y,y') - ik_{y,N}f^i_{\substack{m\\n},N}(y,y')\right)\right)\Bigg|_{y=y_N} = 0 \quad (33)$$

Therefore, variables of the BC at two ends can be obtained:

$$\left\{\begin{array}{l} a_{\substack{m\\n},1} = \left[S^-_{y,i}(y')\right]\left[B_{\substack{m\\n},i}\right]\left[\begin{array}{cc}1 & -i\end{array}\right]^T/\left(k_{y,i}I_{\substack{m\\n},i}\right),\ b_{\substack{m\\n},1} = \left[S^-_{y,i}(y')\right]\left[B_{\substack{m\\n},i}\right]\left[\begin{array}{cc}-i & 1\end{array}\right]^T/\left(k_{y,i}I_{\substack{m\\n},i}\right) \\ c_{\substack{m\\n},N} = \left[S^+_{y,i}(y')\right]\left[A_{\substack{m\\n},i}\right]\left[\begin{array}{cc}1 & -i\end{array}\right]^T/\left(k_{y,i}I_{\substack{m\\n},i}\right),\ d_{\substack{m\\n},N} = \left[S^+_{y,i}(y')\right]\left[A_{\substack{m\\n},i}\right]\left[\begin{array}{cc}-i & 1\end{array}\right]^T/\left(k_{y,i}I_{\substack{m\\n},i}\right)\end{array}\right. \quad (34)$$

where

$$I_{\substack{m\\n},i} = \left[\begin{array}{cc}-i & 1\end{array}\right]\left[B_{\substack{m\\n},i}\right]^T[D_i]\left[A_{\substack{m\\n},i}\right]\left[\begin{array}{cc}-i & 1\end{array}\right]^T \quad (35)$$

Hence, the SGF and the DGF can also be written, according to Equations (20) and (21).

Example: A Rectangular Waveguide Longitudinally Filled with Multi-Layered Media

The rectangular waveguide longitudinally filled with multi-layered media is exhibited in Figure 3, which is stratified along the y-axis. Therefore,

$$g_m(x,z) = \sin(k_x x)\sin(hz), g_n(x,z) = \cos(k_x x)\cos(hz), F_m = 4/(ab), F_n = 4(2-\delta_{k_c,0})/(ab), k_x = p\pi/b, h = q\pi/a \quad (36)$$

where b is the height of the rectangular waveguide at x-axis, and a is the width of the rectangular waveguide at z-axis.

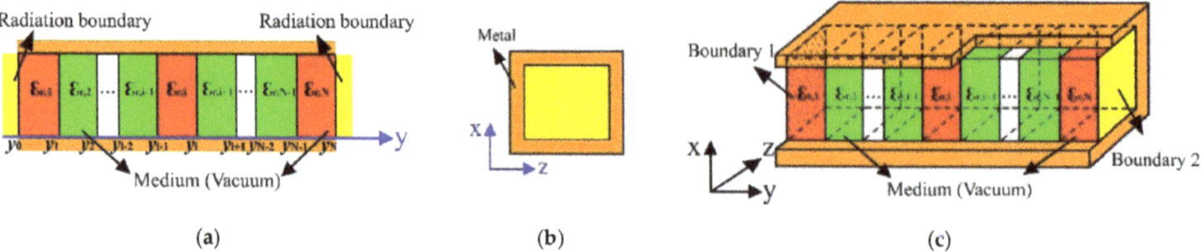

Figure 3. Geometry of the rectangular waveguide longitudinally filled with multi-layered media. (**a**) front view; (**b**) side view; (**c**) 3D view.

3.3. Metal and Infinite Radiation Boundary Conditions

Assuming that boundary 1 is metal and boundary 2 satisfies the infinite radiation condition in Figure 1, then

$$\frac{d}{dy}f^i_{m,1}(y,y')\Big|_{y=y_0}=0,\quad f^i_{n,1}(y,y')\Big|_{y=y_0}=0,\ \lim_{y_N\to\infty}\left(y^{\frac{s-1}{2}}\left(\frac{d}{dy}f^i_{\substack{m\\n},N}(y,y')-ik_{y,N}f^i_{\substack{m\\n},N}(y,y')\right)\right)\Big|_{y=y_N}=0 \tag{37}$$

Therefore,

$$a_{m,1}=0,\quad c_{m,N}=-id_{m,N},\quad b_{n,1}=0,\quad d_{n,N}=ic_{n,N} \tag{38}$$

Equations (24) and (25) can be derived as:

$$\begin{cases} b_{m,1}=\left[S^-_{y,i}(y')\right][B_{m,i}]\begin{bmatrix}-i & 1\end{bmatrix}^T/(k_{y,i}I_{m,i}),d_{m,N}=\left[S^+_{y,i}(y')\right][A_{m,i}]\begin{bmatrix}0 & 1\end{bmatrix}^T/(k_{y,i}I_{m,i}) \\ a_{n,1}=\left[S^-_{y,i}(y')\right][B_{n,i}]\begin{bmatrix}1 & i\end{bmatrix}^T/(k_{y,i}I_{n,i}),c_{n,N}=\left[S^+_{y,i}(y')\right][A_{n,i}]\begin{bmatrix}1 & 0\end{bmatrix}^T/(k_{y,i}I_{n,i}) \end{cases} \tag{39}$$

where

$$I_{m,i}=\begin{bmatrix}-i & 1\end{bmatrix}[B_{m,i}]^T[D_i][A_{m,i}]\begin{bmatrix}0 & 1\end{bmatrix}^T,I_{n,i}=\begin{bmatrix}1 & i\end{bmatrix}[B_{n,i}]^T[D_i][A_{n,i}]\begin{bmatrix}1 & 0\end{bmatrix}^T \tag{40}$$

The undetermined variables have been derived, and the corresponding SGF and DGF can be written, according to Equations (20) and (21).

4. Application on the Dispersion Analysis for a Meander Line Slow-Wave Structure

As shown in Figure 4, the meander line slow-wave structure (ML-SWS) is composed of a meander line and a dielectric loaded waveguide. Here, the meander line is clamped with dielectrics, whose relative permittivity is ε_r. For easy calculation, the thickness of the meander line is regarded as zero here. The parameters of the ML-SWS are given in Table 1.

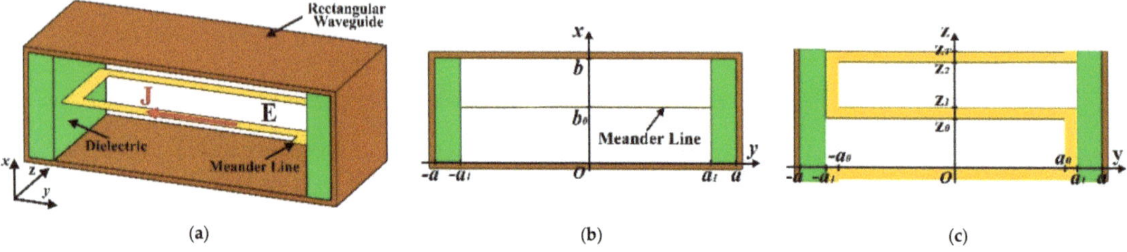

Figure 4. The photograph of a meander line slow-wave structure: (**a**) 3D model view; (**b**) front view; (**c**) top view.

Table 1. Parameters of ML-SWS (size dimensions in millimeters).

b	b_0	A	a_0	a_1	z_0	z_1	z_2	z_T	ε_r
1	0.5	0.7	0.45	0.5	0.2	0.25	0.45	0.5	2

According to the metal boundary condition, the tangential electric field $\vec{E}_{\parallel}$ on the surface of the meander line should be zero.

$$\vec{E}_{\parallel}\left(\vec{R}\right) = 0, \vec{R} \in \text{Meander Line} \tag{41}$$

According to [9], the electric field $\vec{E}\left(\vec{R}\right)$ can be written as:

$$\vec{E}\left(\vec{R}\right) = i\omega\mu \int\limits_{V'} \vec{\vec{G}}\left(\vec{R},\vec{R'}\right)\cdot\vec{J}\left(\vec{R'}\right)dV' \tag{42}$$

where $\vec{\vec{G}}$ is the DGF of multi-layered plane media, which can be obtained as mentioned before. $\vec{J}$ is a current source along the meander line. Here, the current source can be expanded with sets of Ritz basic functions $\varphi_{D,s}$ such as the electric fields [27]:

$$J_{\rm D} = \sum_{s=0}^{\infty} A_{\rm D,s}\varphi_{\rm D,s} \tag{43}$$

where subscript "D" represents the direction of the current J and $A_{\rm D,s}$ is a coefficient.

Bringing Equations (42) and (43) into Equation (41), the electromagnetic field expressions can be obtained. The expressions includes three sets of $\{A_{\rm D,s}\}$, position vector $\vec{R}$ and $\{\omega,\theta\}$, where ω is wave frequency and θ is the phase shift in a period.

Moreover, the cross product of Equation (41) with Ritz basic functions can be expressed in matrix form [28]:

$$[Y][A] = 0 \tag{44}$$

As a result, the dispersion function of the ML-SWS can be obtained as:

$$|Y| = 0 \tag{45}$$

The dispersion characteristic of the ML-SWS can be obtained by Equation (45), where the upper range of both parameters s and t are 0 to 2. This calculation procedure can be performed in 2.4 min, which is only one-quarter of the time cost by HFSS code.

Based on the derived DGF of multi-layered plane media, the dispersion characteristics of the ML-SWS can be obtained, as shown in Figure 5. The obtained theoretical results are also compared with the simulated results from HFSS code. Furthermore, the relative error between them is under 3%, which is marked with a dashed line.

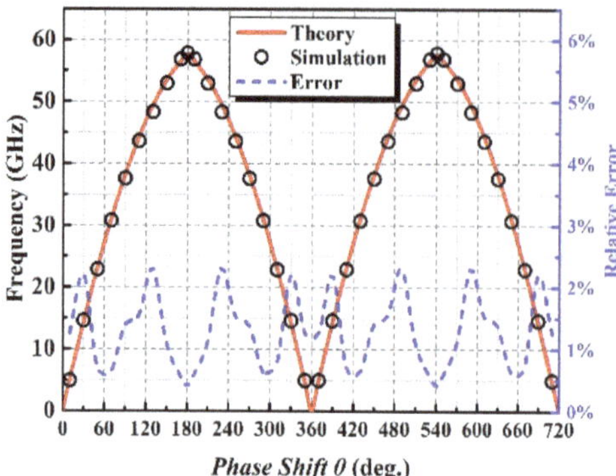

Figure 5. Dispersion characteristics and comparison of the ML-SWS.

Moreover, the effect from the relative permittivity (ε_r) of the dielectric is also studied, as shown in Figure 6. The upper cut-off frequency of the mode with the larger relative permittivity is smaller than that with the lower one. With the relative permittivity increasing, phase velocity decreases accordingly. The relative error between the theoretical results and simulated results from HFSS code are also within 3%, which indicates the theoretical results and simulated results with different relative permittivity are in good agreement.

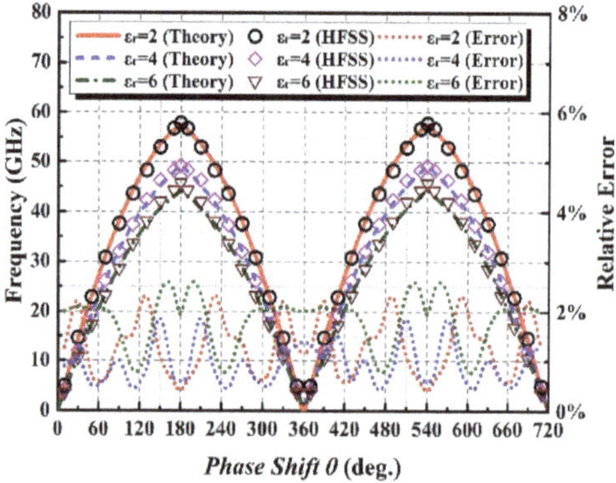

Figure 6. Dispersion of a ML-SWS with different relative permittivity ε_r of two dielectrics.

5. Conclusions

A method for obtaining the Green's functions of multi-layered plane media has been proposed in this paper. In this method, the Green's functions consist of the generation functions. In the process of deducing the generation functions, the boundary conditions between the adjacent layers and at two ends have been considered, which are represented by a series of separable variables. Because these variables are separated in matrix form, the corresponding boundary equations can be independent. Consequently, the form of the generation functions and the obtained Green's functions can be explicit and can be in consistence of different boundary conditions, and the corresponding expression can be expressed with the more friendly computer code.

Moreover, both the dyadic Green's functions (DGF) and scalar Green's functions (SGF) have been obtained. The obtained results are in good agreement with the predecessors' works.

Furthermore, as the application, a ML-SWS with different relative permittivity has been analyzed. The calculation procedure can be performed in 2.4 min, which is only one-quarter of the time cost by HFSS code. The relative error between the theoretical results and the simulated ones with different relative permittivity is under 3%, which demonstrates that the proposed DGF can be suitable for electromagnetic analysis of complicated structures, including the ML-SWS.

Author Contributions: Conceptualization, Z.W., J.L. and W.L.; methodology, Z.W.; software, Z.W.; validation, Z.W. and W.L.; formal analysis, Z.W.; investigation, Z.W.; resources, Z.W.; data curation, Z.W.; writing—original draft preparation, Z.W.; writing—review and editing, Z.W. and J.L.; visualization, Z.W.; supervision, Z.W.; project administration, J.L. All authors have read and agreed to the published version of the manuscript.

Funding: This research received no external funding

Conflicts of Interest: The authors declare no conflict of interest.

Appendix A

The matrixes $[T_{m,k}]$, $[T_{n,k}]$, $[R_{m,l}]$, $[R_{n,l}]$ and $[D_i]$ are defined as follows:

$$[T_{m,k}] = \begin{bmatrix} \frac{\sqrt{\varepsilon_{r,k+1}}}{\sqrt{\varepsilon_{r,k}}}\frac{k_{y,k}}{k_{y,k+1}}\cos\Big(k_{y,k}(y_k - y_{k-1})\Big) & -\frac{\sqrt{\varepsilon_{r,k+1}}}{\sqrt{\varepsilon_{r,k}}}\frac{k_{y,k}}{k_{y,k+1}}\sin\Big(k_{y,k}(y_k - y_{k-1})\Big) \\ \frac{\sqrt{\varepsilon_{r,k}}}{\sqrt{\varepsilon_{r,k+1}}}\sin\Big(k_{y,k}(y_k - y_{k-1})\Big) & \frac{\sqrt{\varepsilon_{r,k}}}{\sqrt{\varepsilon_{r,k+1}}}\cos\Big(k_{y,k}(y_k - y_{k-1})\Big) \end{bmatrix}, [T_{m,0}] = \begin{bmatrix} 1 & 0 \\ 0 & 1 \end{bmatrix} \quad \text{(A1)}$$

$$[R_{m,l}] = \begin{bmatrix} \frac{\sqrt{\varepsilon_{r,l}}}{\sqrt{\varepsilon_{r,l+1}}}\frac{k_{y,l+1}}{k_{y,l}}\cos\Big(k_{y,l+1}(y_{l+1} - y_l)\Big) & -\frac{\sqrt{\varepsilon_{r,l}}}{\sqrt{\varepsilon_{r,l+1}}}\frac{k_{y,l+1}}{k_{y,l}}\sin\Big(k_{y,l+1}(y_{l+1} - y_l)\Big) \\ \frac{\sqrt{\varepsilon_{r,l+1}}}{\sqrt{\varepsilon_{r,l}}}\sin\Big(k_{y,l+1}(y_{l+1} - y_l)\Big) & \frac{\sqrt{\varepsilon_{r,l+1}}}{\sqrt{\varepsilon_{r,l}}}\cos\Big(k_{y,l+1}(y_{l+1} - y_l)\Big) \end{bmatrix}, [R_{m,N}] = \begin{bmatrix} 1 & 0 \\ 0 & 1 \end{bmatrix} \quad \text{(A2)}$$

$$[T_{n,k}] = \begin{bmatrix} \frac{k_{y,k}}{k_{y,k+1}}\cos\Big(k_{y,k}(y_k - y_{k-1})\Big) & -\frac{k_{y,k}}{k_{y,k+1}}\sin\Big(k_{y,k}(y_k - y_{k-1})\Big) \\ \sin\Big(k_{y,k}(y_k - y_{k-1})\Big) & \cos\Big(k_{y,k}(y_k - y_{k-1})\Big) \end{bmatrix}, [T_{n,0}] = \begin{bmatrix} 1 & 0 \\ 0 & 1 \end{bmatrix} \quad \text{(A3)}$$

$$[R_{n,l}] = \begin{bmatrix} \frac{k_{y,l+1}}{k_{y,l}}\cos\Big(k_{y,l+1}(y_{l+1} - y_l)\Big) & -\frac{k_{y,l+1}}{k_{y,l}}\sin\Big(k_{y,l+1}(y_{l+1} - y_l)\Big) \\ \sin\Big(k_{y,l+1}(y_{l+1} - y_l)\Big) & \cos\Big(k_{y,l+1}(y_{l+1} - y_l)\Big) \end{bmatrix}, [R_{n,N}] = \begin{bmatrix} 1 & 0 \\ 0 & 1 \end{bmatrix} \quad \text{(A4)}$$

$$[D_i] = \begin{bmatrix} \sin(k_{y,i}(y_i - y_{i-1})) & \cos(k_{y,i}(y_i - y_{i-1})) \\ \cos(k_{y,i}(y_i - y_{i-1})) & -\sin(k_{y,i}(y_i - y_{i-1})) \end{bmatrix} \quad \text{(A5)}$$

References

1. Pincherle, L. Electromagnetic Wave in Metal Tubes Filled Longitudinally with Two Dielectrics. *Phys. Rev.* **1944**, *55*, 118–130. [CrossRef]
2. Eroglu, A.; Lee, Y.H.; Lee, J.K. Dyadic Green's functions for multi-layered uniaxially anisotropic media with arbitrarily oriented optic axes. *IET Microw. Antennas Propag.* **2011**, *5*, 1779–1788. [CrossRef]
3. Kourkoulos, V.N.; Cangellaris, A.C. Accurate approximation of Green's functions in planar stratified media in terms of a finite sum of spherical and cylindrical waves. *IEEE Trans. Antennas Propag.* **2006**, *54*, 1568–1576. [CrossRef]
4. Boix, R.R.; Mesa, F.; Medina, F. Application of Total Least Squares to the Derivation of Closed-Form Green's Functions for Planar Layered Media. *IEEE Trans. Microw. Theory Tech.* **2007**, *55*, 268–280. [CrossRef]
5. Wu, B.P.; Tsang, L. Fast Computation of Layered Medium Green's Functions of Multilayers and Lossy Media Using Fast All-Modes Method and Numerical Modified Steepest Descent Path Method. *IEEE Trans. Microw. Theory Tech.* **2008**, *56*, 1446–1454. [CrossRef]
6. Kwon, M.S. A Numerically Stable Analysis Method for Complex Multilayer Waveguides Based on Modified Transfer-Matrix Equations. *IEEE J. Lightwave Technol.* **2009**, *27*, 4407–4414. [CrossRef]
7. Tai, C.T. Dyadic Green's Functions for a Rectangular Waveguide Filled with Two Dielectrics. *J. Electromagn. Waves Applicat.* **1988**, *2*, 245–253. [CrossRef]
8. Chew, W. *Waves and Field in Inhomogeneous Media*; IEEE Press Series on Electromagnetic Wave Theory; IEEE: New York, NY, USA, 1999.

9. Tai, C.T. *Dyadic Green Functions in Electromagnetic Theory*, 2nd ed.; IEEE Press: Piscataway, NJ, USA, 1993.
10. Felsen, L.B.; Marcuvitz, N. *Radiation and Scattering of Waves*; IEEE Press Series on Electromagnetic Wave Theory; IEEE Press: New York, NY, USA, 1994.
11. Jin, H.; Lin, W. Dyadic Green's functions for a rectangular waveguide with an E-plane dielectric slab. *IEE Proc. H Microw. Antennas Propag.* **1990**, *137*, 231–234. [CrossRef]
12. Joubert, J.; McNamara, D.A. Dyadic Green's function of electric type for inhomogeneously loaded rectangular waveguides. *IEE Proc. H Microw. Antennas Propag.* **1989**, *136*, 469–474. [CrossRef]
13. Hanson, G.W. Dyadic Green's function for a multi-layered planar medium—A dyadic eigenfunction approach. *IEEE Trans. Antennas Propag.* **2004**, *52*, 3350–3356. [CrossRef]
14. Hanson, G.W. Dyadic Eigenfunctions and Natural Modes for Hybrid Waves in Planar Media. *IEEE Trans. Antennas Propag.* **2004**, *52*, 941–947. [CrossRef]
15. Qiu, C.W.; Yao, H.Y.; Li, L.W.; Zouhdi, S.; Yeo, T.S. Eigenfunctional representation of dyadic Green's functions in multilayered gyrotropic chiral media. *J. Phys. A Math. Theor.* **2007**, *40*, 5751–5766. [CrossRef]
16. Lobo, A.E.; Tsoy, E.N.; Martijn de Sterke, C. Green function method for nonlinear elastic waves in layered media. *J. Appl. Phys.* **2001**, *90*, 3762–3770. [CrossRef]
17. How, H.; Zuo, X.; Vittoria, C. Dyadic Green's function calculations on a layered dielectric/ferrite structure. *J. Appl. Phys.* **2001**, *89*, 6722–6724. [CrossRef]
18. Song, W.M. *Dyadic Green's Function and Operator Theory of Electromagnetic Filed*; Press of University of Science and Technology of China: Hefei, China, 1991.
19. Sphicopoulos, T.; Teodoridis, V.; Gardiol, F.E. Dyadic green function for the electromagnetic field in multilayered isotropic media: An operator approach. *IEE Proc. H Microw. Antennas Propag.* **1985**, *132*, 329–334. [CrossRef]
20. Chen YP, P.; Chew, W.C.; Jiang, L.J. A New Green's Function Formulation for Modeling Homogenerous Objects in Layered Medium. *IEEE Trans. Antennas Propag.* **2012**, *60*, 4766–4776. [CrossRef]
21. Jin, H.; Lin, W.; Lin, Y. Dyadic Green's functions for rectangular waveguide filled with longitudinally meltilayered isotropic dielectric and their application. *IEE Proc. Microw. Antennas Propag.* **1994**, *141*, 504–508. [CrossRef]
22. Michalski, K.A.; Mosig, J.R. Multilayered Media Green's Functions in Integral Equation Formulations. *IEEE Trans. Antennas Propag.* **1997**, *45*, 508–519. [CrossRef]
23. Kakade, A.B.; Ghosh, B. Analysis of the rectangular waveguide slot coupled multilayer hemispherical dielectric resonator antenna. *IET Microw. Antennas Propag.* **2012**, *6*, 338–347. [CrossRef]
24. Muhlschlegel, P.; Eisler, H.J.; Martin, O.J.F.; Hecht, B.; Pohl, D.W. Resonant Optical Antennas. *Science* **2005**, *308*, 1607–1609. [CrossRef]
25. Zhu, H.T.; Xue, Q.; Liao, S.W.; Pang, S.W.; Chiu, L.; Tang, Q.Y.; Zhao, X.H. Low-Cost Narrowed Dielectric Microstrip Line-A Three-Layer Dielectric Waveguide Using PCB Technology for Millimeter-Wave Applications. *IEEE Trans. Microw. Theory Tech.* **2017**, *65*, 119–127. [CrossRef]
26. Dey, U.; Hesselbarth, J. Building Blocks for a Millimetere-wave Multiport Multicast Chip-to-Chip Interconnect Based on Dielectric Waveguides. *IEEE Trans. Microw. Theory Tech.* **2018**, *66*, 5508–5520. [CrossRef]
27. Su, Q.C.; Wu, H.S. Green's function solution of the waveguide with two pairs of double ridges. *Acta Electron. Sin.* **1983**, *11*, 81–87. (In Chinese)
28. Wen, Z.; Fan, Y.; Yang, C.; Luo, J.R.; Zhu, F.; Zhu, M.; Guo, W.; Gong, Y.B.; Feng, J.J. Theory, Simulation and Analysis of the High Frequency Characteristics for a Meander Line Slow-wave Structure Based on Field-matching Methods with Dyadic Green's Function. *IEEE Trans. Electron Devices* **2020**, *67*, 697–703. [CrossRef]

MDPI

.rticle

Inverse Design of a Microstrip Meander Line Slow Wave Structure with XGBoost and Neural Network

Yijun Zhu, Yang Xie, Ningfeng Bai * and Xiaohan Sun

Research Center for Electronic Device and System Reliability, Southeast University, Nanjing 210096, China; 220191386@seu.edu.cn (Y.Z.); xieyang@seu.edu.cn (Y.X.); xhsun@seu.edu.cn (X.S.)
* Correspondence: bnfeng@seu.edu.cn

Abstract: We present a new machine learning (ML) deep learning (DL) synthesis algorithm for the design of a microstrip meander line (MML) slow wave structure (SWS). Exact numerical simulation data are used in the training of our network as a form of supervised learning. The learning results show that the training mean squared error is as low as 5.23×10^{-2} when using 900 sets of data. When the desired performance is reached, workable geometry parameters can be obtained by this algorithm. A D-band MML SWS with 20 GHz bandwidth at 160 GHz center frequency is then designed using the auto-design neural network (ADNN). A cold test shows that its phase velocity varies by 0.005 c, and the transmission rate of a 50-period SWS is greater than −5 dB with the reflectivity below −15 dB when the frequency is from 150 to 170 GHz. Particle-in-cell (PIC) simulation also illustrates that a maximum power of 3.2 W is reached at 160 GHz with 34.66 dB gain and output power greater than 1 W from 152 to 168 GHz.

Keywords: deep learning (DL); machine learning (ML); microstrip meander line slow wave structure (MML-SWS); D-band

Citation: Zhu, Y.; Xie, Y.; Bai, N.; Sun, X. Inverse Design of a Microstrip Meander Line Slow Wave Structure with XGBoost and Neural Network. *Electronics* **2021**, *10*, 2430. https://doi.org/10.3390/electronics10192430

Academic Editor: Geok Ing Ng

Received: 17 August 2021
Accepted: 5 October 2021
Published: 7 October 2021

1. Introduction

High-frequency millimeter-wave communication is receiving increasing attention due to the development of 6G and next-generation communication networks [1]. It is difficult to realize high-power millimeter-wave sources over the W-band because of the power limitation of solid-state devices [2]. As a core component of high-power microwave devices, the traveling wave tube (TWT) has a wide range of applications in millimeter-wave fields, and has been applied in terahertz wave transmission systems [3]. Compared with solid-state devices, TWTs have obvious advantages at high frequencies [4]. Therefore, it is necessary to carry out research on high-frequency TWT technology to promote the development of high-frequency millimeter-wave technology [5]. Among various types of TWTs, spiral and folded waveguides are the most common slow wave structure (SWS) in the microwave and millimeter-wave bands [6]. In [7], a double corrugated waveguide (DCW) SWS was designed to support a beam voltage of 13 kV with a wide bandwidth of about 20 GHz, obtaining an interaction impedance of approximately 1.5 Ω at D band. An SWS for a W-band folded-waveguide TWT with an operating bandwidth of around 3 GHz was also designed, delivering an output power of 50 W at the operating voltage of 13.5 kV and operating beam current of 80 mA. Obviously, these SWSs require high voltages to provide a high output power, which is the obstacle preventing the application of vacuum electronic device (VED) in modern communication systems.

Planar microstrip meander line (MML) SWSs are more conducive to microfabrication techniques, which can offer simple construction, wide bandwidth and low operation voltage [8]. MML SWSs have demonstrated a low voltage of 3–5 kV at V-band and W-band [9,10]. Recently, Zhen et al. used a concentric arc MML SWS to obtain 44 W output power with 18.6 dB gain working at 720 V [11].

Meanwhile, deep learning (DL) [12,13], a supervised method of pattern analysis with a multilayered structure, has been widely applied in inverse-design over the past decade [14,15]. In current research of inverse-design based on DL, deep neural networks (DNN) are trained using tensors encoded with structure and spectrum characteristics [16,17], demonstrating the use of DL technology as an inverse-design tool in microwave and optical wave fields. In [18], a purpose-designed DL architecture made up of a convolutional neural network (CNN) and a fully-connected neural network (FCNN) was used to automatically model and optimize three-dimensional chiral metamaterials, achieving high numerical accuracy in plasmonic meta-surfaces. The method realized the design-on-demand function and produced suitable meta-atom geometric parameters to fulfill the given requirements. A typical multi-layer FCNN was also successfully applied to solve effective refractive indices of the fundamental waveguide mode in a silicon nitride channel waveguide for both polarizations of light [19]. The DL model was only trained with sixteen data points and could accurately predict patterns in the effective refractive indices. Malkiel et al. introduced a DL architecture that was applied to the design and characterization of metal-dielectric sub-wavelength nanoparticles. Their approach of training a bidirectional network that goes from the optical response spectrum to the nanoparticle geometry and back was significantly more effective than the alternative method of training separate models for design and characterization tasks [20]. This data-driven technique has been applied to tackle challenging problems in a wide range of fields. Applying DL to achieve practical parameters for an SWS is a highly effective method for their design.

While these DL algorithms have very powerful learning and prediction capabilities, their interpretability remains a significant challenge. Especially in the field of VED, where the dimensions of device structure and spectrum are low and the data set is limited, the algorithm could provide sufficient design guidance rather than simply be used as a blackbox function.

This paper aims to rapidly design an MML SWS according to the target center frequency and bandwidth using our proposed algorithm. An XGBoost-DNN composite structure [21] is applied to inverse design a practical MML SWS. The optimized parameters of an MML SWS are then obtained using supervised machine learning algorithms according to the desired bandwidth and center frequency of the MML SWS. The mean squared error (MSE) is reduced to 0.001 using 900 groups of data. An MML SWS is then designed for particle-in –cell (PIC) simulations using the parameters obtained by this method, and the results show that the obtained parameters work well for the MML SWS.

2. MML Structure and Cold Parameters

The unit structure of the proposed MML with a metal shield consists of two parts, the dielectric substrate, and the MML, which are shown in Figure 1. The dielectric substrate material is silicon dioxide (SiO_2), the thickness of the dielectric substrate is set as h, the relative dielectric constant ε is 3.75, and the tangent loss, tanδ, is 0.0004. The MML is pure copper, with electrical conductivity at 2e7 S/m. Its thickness is $t = 0.01$ mm, with a line width of w, a distance between two adjacent transverse microstrip lines of s, and a transverse length of l. H is the height between the MML and metal shield and is fixed at 0.75 mm; L is the transverse width of the metal shield, where L = 2 * ls + l and ls is fixed at 0.2 mm.

The main characterizations of the MML are phase velocity and transmission, which determine the performance of an MML SWS. Therefore, the phase velocity and transmission versus frequency are simulated using CST studio suite. In order to describe the spectrum characteristics of the MML, the characteristic parameters v, d, $Smin$ and $Smax$ from the spectrum shown in Figure 2a,b are defined, where v is normalized phase velocity at the central frequency, d is the difference between the phase velocity at the lower bandwidth and that at the upper bandwidth, which represents the flatness of phase velocity within the required bandwidth.

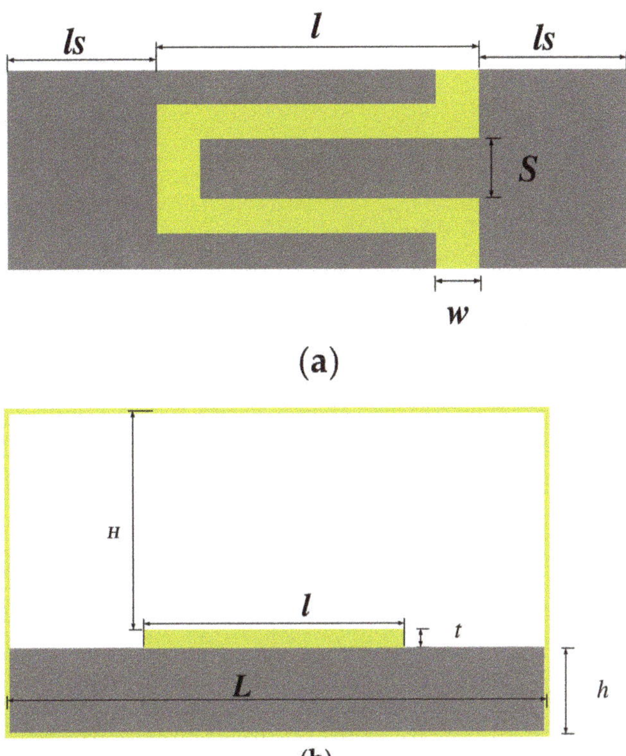

Figure 1. (**a**) Front view; and (**b**) top view of the MML unit structure.

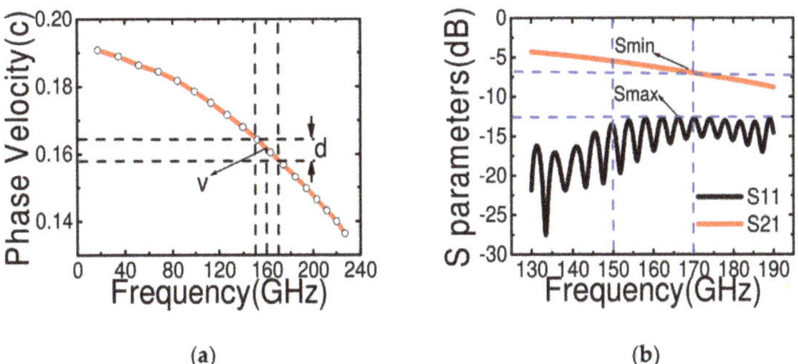

(**a**) (**b**)

Figure 2. Cold-test characteristic of the MML structure: (**a**) Dispersion characteristic; (**b**) Transmission (S21) and reflection (S11).

Figure 2b shows the transmission characteristics of the MML, where the minimum transmission (*Smin*) and the maximum reflection (*Smax*) in the bandwidth are defined. These two parameters represent the transmission performance of the MML structure.

3. Method

XGBoost is trained to learn the characteristic of MML and inverse design the parameters using the four characteristic parameters v, d, *Smin* and *Smax*. The software builds a one-way mapping between the structural and spectral parameters. XGBoost provides similar functionality to that of CST, but has a much faster calculation speed after training with a dataset simulated by CST. It also provides the importance index of structural parameters to the optimization of different spectrum parameters. DNN plays the exact opposite role to XGBoost, offering different structural parameters to the pre-trained XGBoost, which then predicts the spectral parameters of the structural parameters according to the previously established one-way mapping relationship. This process is called inverse design.

XGBoost uses the greedy algorithm to traverse all possible values of the four parameters (s, l, w, h) of the MML structure and calculate the importance index. It then continuously splits the data set according to the importance index to construct a decision tree.

As an integration model of decision trees, the output of XGBoost is the weighted sum of the outputs of the k decision trees. The best split in each tree learning must be determined. In order to do so, a split finding algorithm considers all possible splits on all four structural features, which is called the exact greedy algorithm [21]. The objective function at step t of XGBoost is:

$$L^{(t)} = \sum_{i=1}^{n}\left[g_i f_t(x_i) + \frac{1}{2}h_i f_t^2(x_i)\right] + \Omega(f_t)$$

where g_i and h_i are the first and second-order gradient statistics on the loss function (MSE), and $\Omega(f_t) = \gamma T$ is the regularization parameter used to solve the problem of over-fitting by limiting the size of each decision tree output value.

The proposed microstrip meander line predict model (MMLPM) is schematically depicted in Figure 3a. In the learning process of XGBoost, the importance index for all features is calculated according to the formula of the corresponding optimal value $w_j = -\frac{\sum_{i\in I_j} g_i}{\sum_{i\in I_j} h_i+\lambda}$ [21], where λ is an artificially defined hyperparameter used to control the weight of the regularization parameters, and subscript i represents the set, the data are present according to the split point. Moreover, these indicators will provide a meaningful reference for designers to configure the MML. As shown in Figure 3b, the influence of the metal folding line length on the phase velocity v and transmission $Smin$ is greater than other structural parameters. In the same way, adjusting the value pairs of s is more helpful to improve the performance of d and $Smin$.

In this paper, 900 samples were simulated by CST to train and validate XGBoost. Based on the results of several previous simulations, s was set to be in the range from 0.01 to 0.03 mm, l was set to be in the range from 0.1 to 0.3 mm, h was set to be in the range from 0.015 to 0.04 mm, w was set to be in the range from 0.015 to 0.04 mm. The entire data set had a total of 900 data points for network learning and verification at ratios of 0.7 and 0.3, respectively.

We use XGBoost to establish the forward mapping relationship between the structural and spectral parameters, which is more interpretable than neural networks. However, our goal is to use deep learning and machine learning to reverse design SWS rather than simply predict the spectrum. Therefore, we also train a fully connected forward neural network (FNN) whose input is a set of spectral parameters and output is a set of structural parameters. Adam optimization algorithm is used [22], which can automatically adapt to adjust the learning rate. As shown in Figure 4a, the basic structure of a DNN consists of three components: An input layer, a hidden layer, and an output layer. These layers are an FCNN, meaning every neuron in one layer is connected to all neurons in the previous layer. Therefore, the output of neurons in the previous layer is the input of neurons in the next layer, and each connection has a weighted value w. In the equations in Figure 4b, σ is the activation function. The goal of each iteration is to update these weights so that the prediction results are increasingly similar to the simulation data. There is no connection between neurons within the same layer. In the learning process of a neural network, losses in learning are propagated backward, and can be measured by the MSE or linear errors.

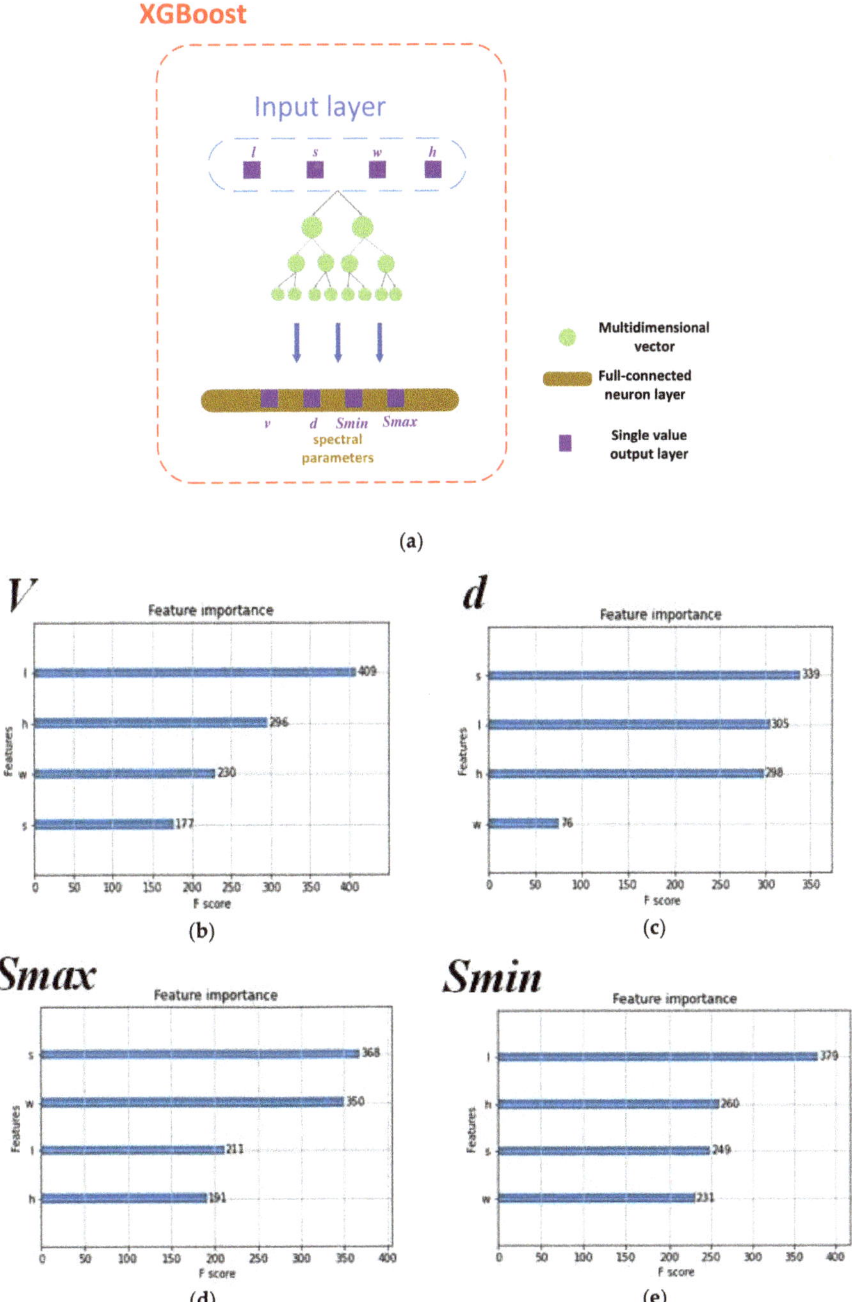

Figure 3. (**a**) MMLPM basic structure; (**b**) the corresponding optimal value with phase velocity; (**c**) the corresponding optimal value with phase velocity flatness; (**d**) the corresponding optimal value with *Smax;* (**e**) the corresponding optimal value with *Smin*.

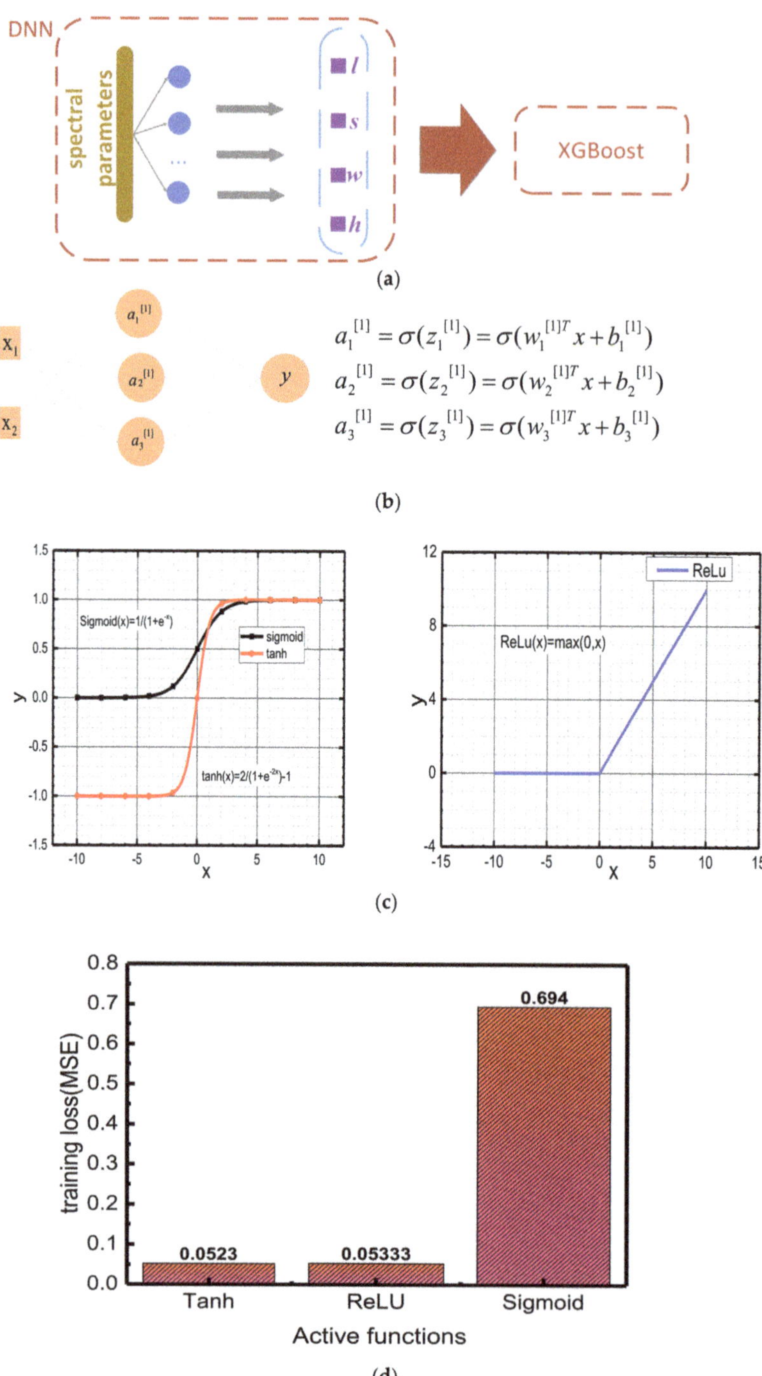

Figure 4. (**a**) XGBoost-DNN composite structure; (**b**) basic structure of a fully connected deep neural network; (**c**) three different active functions; (**d**) training loss with different active functions.

For a node in a hidden layer of a neural network, the calculation of its activation value is divided into two steps: (1) The values of the nodes x_1 and x_2 are given when entering the hidden node to achieve a linear transformation, and the value of $Z^{[1]} = w^1x_1 + w^2x_2 + b^{[1]} = w^{[1]}x + b^{[1]}$ is calculated, where superscript 1 designates the first hidden layer. (2) For a nonlinear transformation; that is, a nonlinear activation function, the output of the node $a(1) = g(z(1))$ is caculated, where $g(z)$ is a nonlinear function. Graphs of three of the most commonly used active functions are provided in Figure 4b. All three active functions in our method are tested, and the training loss is shown in Figure 4c. The backpropagation loss of the FNN is calculated from the mean square error function for the output of XGBoost and the input of the FNN [23], which means the FNN is trained to offer XGBoost a suitable set of structural parameters. Once the training is done, the corresponding structure parameters can be obtained by inputting the target spectral parameters to the FNN. Obviously, the results of Tanh and ReLu have larger gradients than that of sigmoid near the center value, leading to faster weight update speed of multi-layer neural network. Although the training error of tanh activation function is minimized, the training losses of ReLu and tanh are very close to each other. Moreover, ReLu can effectively avoid the problem of gradient disappearance in DNN [23]. Therefore, ReLu function is chosen as the active function in our method.

4. Results and Discussion

To validate the XGBoost-DNN, a range of initialization spectral properties were offered to the model, where v was less than 0.15c, d was less than 0.005, $Smax$ was less than −5 dB, $Smin$ was higher than −5 dB. For an SWS with the desired center frequency of 160 GHz and bandwidth of 20 GHz, a set of the specific structural parameters designed by XGBoost-DNN were given as: s = 0.012 mm, l = 0.2 mm, h = 0.02 mm, w = 0.016 mm.

The cold-test and transmission characteristics are shown in Figure 5a,b. As shown in the figure, the maximum reflection of our optimized structure remains below −15 dB from 150 to170 GHz, the phase velocity is 0.134c at the central frequency, and the on-axis coupled pierce impedance at 0.03 mm above the metal MML is 14.3 Ω, which means the optimized structure performs as expected.

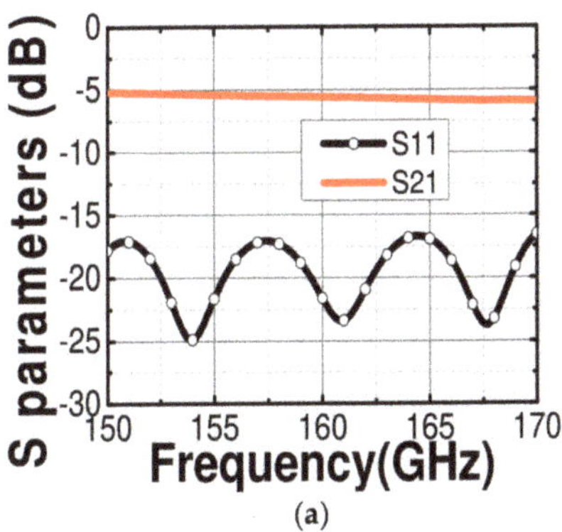

(a)

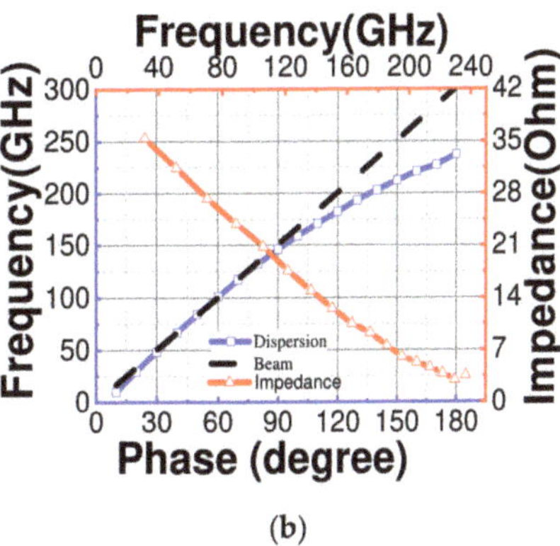

(b)

Figure 5. (**a**) Transmission characteristics from 150 to 170 GHz of the designed MML-SWS; (**b**) dispersion characteristic and coupled impedance of the structure.

PIC simulation was performed to validate the designed MML-SWS. A 4.59-kV operating voltage and a 0.232 mm × 0.02 mm sheet beam with 50 mA current were applied with a 0.6 T longitudinal magnetic field in the PIC simulation. The schematic model in CST is shown in Figure 6a. In addition, according to the guidance of previous literature [24], three section attenuators with a maximum tangent loss of 0.24 were used, and each attenuator had a length of 15 periods, located at periods 25, 65 and 105, respectively. As shown in

Figure 6b, when the input signal power is 1 mW, the stable output power with 180 periods reaches 3 W with 36.66 dB gain at 160 GHz. Meanwhile, the power gain is above 30 dB over a 20 GHz bandwidth range from 150 to 170 GHz with input power at 1 mW, as shown in Figure 6c, which validates that the design geometry meets our requirements.

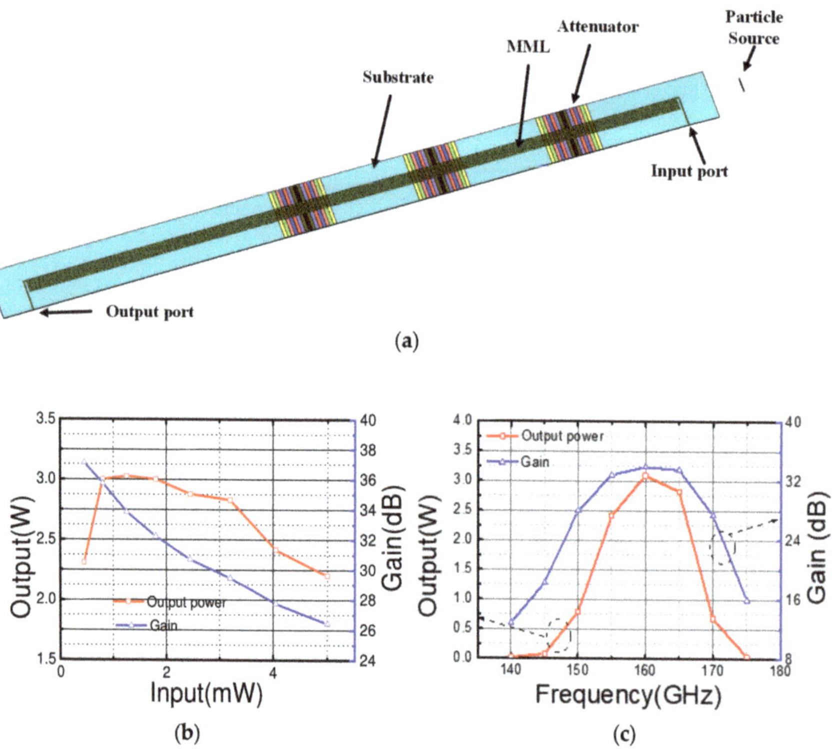

Figure 6. (**a**)Schematic model for PIC simulation; (**b**) output power and gain versus input signal power; (**c**) output power and gain versus frequency.

5. Conclusions

We successfully utilized XGBoost and DNN technology to design an MML-SWS for the optimization of a cold-test and transmission characteristics in this work. The raw data collection was based on the simulation results obtained from CST with four observed parameters of phase velocity at the center frequency, dispersion flatness, minimal transitivity, and maximal reflectivity between the bandwidth. The XGBoost-DNN could learn these relations from the raw data to determine the optimal parameters. Once the center frequency and bandwidth were given, the appropriate SWS could be designed automatically.

In this paper, only D-band parameters were learned by our algorithm. In future work, the learning database will extend the frequency range from the Ka-band to the G-band. Meanwhile, other types of SWSs also could use this methodology to provide a fast and creative technique to research vacuum electronic devices.

Author Contributions: Conceptualization, N.B., Y.Z. and X.S.; methodology, Y.Z. and N.B.; software, Y.Z.; validation, Y.Z. and Y.X.; investigation, Y.Z. and Y.X.; resources, Y.Z. and Y.X.; data curation, Y.Z. and Y.X.; writing—original draft preparation, Y.Z. and Y.X.; writing—review and editing, N.B. and X.S.; supervision, N.B. and X.S.; project administration, N.B.; funding acquisition, N.B. All authors have read and agreed to the published version of the manuscript.

Funding: This research was funded by the National Natural Science Foundation of China, grant number 61871110. The APC was funded by Southeast University.

Conflicts of Interest: The authors declare no conflict of interest.

References

1. Lee, Y.L.; Qin, D.; Wang, L.C.; Sim, G.H. 6G Massive Radio Access Networks: Key Applications, Requirements and Challenges. *IEEE Open J. Veh. Technol.* **2021**, *2*, 54–66. [CrossRef]
2. Shiffler, D.; Nation, J.A.; Schachter, L.; Ivers, J.D.; Kerslick, G.S. Review of high power traveling wave tube amplifiers. In Proceedings of the IEEE MTT-S Microwave Symposium Digest, Albuquerque, NM, USA, 1–5 June 1992.
3. Gong, Y.B.; Zhou, Q.; Hu, M.; Zhang, Y.H.; Li, X.Y.; Gong, H.R.; Wang, J.X.; Liu, D.W.; Liu, Y.H.; Duan, Z.Y.; et al. Some Advances in Theory and Experiment of High-Frequency Vacuum Electron Devices in China. *IEEE Trans. Plasma Sci.* **2019**, *47*, 1971–1990. [CrossRef]
4. Ryskin, N.M.; Torgashov, R.A.; Starodubov, A.V.; Rozhnev, A.G.; Serdobintsev, A.A.; Pavlov, A.M.; Galushka, V.V.; Bessonov, D.A.; Ulisse, G.; Krozer, V. Development of microfabricated planar slow-wave structures on dielectric substrates for miniaturized millimeter-band traveling-wave tubes. *J. Vac. Sci. Technol. B* **2021**, *39*, 013204. [CrossRef]
5. Paoloni, C.; Gamzina, D.; Letizia, R.; Zheng, Y.; Luhmann, N.C., Jr. Millimeter wave traveling wave tubes for the 21st Century. *J. Electromagn. Waves Appl.* **2021**, *35*, 567–603. [CrossRef]
6. Billa, R.B.L.R.; Rao, J.M.; Letizia, R.; Paoloni, C. Design of D-band Double Corrugated Waveguide TWT for Wireless Communications. In Proceedings of the 2019 International Vacuum Electronics Conference (IVEC), Busan, Korea, 28 April–2 May 2019.
7. Sumathy, M.; Datta, S.K. Design and Characterization of a W-Band Folded-Waveguide Slow-Wave Structure. *J. Infrared Milli Terahz. Waves* **2017**, *38*, 538–547. [CrossRef]
8. Starodubov, V.; Serdobintsev, A.A.; Pavlov, A.M.; Galushka, V.V.; Mitin, D.M.; Ryskin, N.M. A novel microfabrication technology of planar microstrip slow-wave structures for millimeter-band traveling-wave tubes. In Proceedings of the IEEE International Vacuum Electronics Conference (IVEC), Monterey, CA, USA, 24–28 April 2018; pp. 333–334.
9. Ryskin, N.M.; Rozhney, A.G.; Starodubov, A.V.; Serdobintsev, A.A.; Pavlov, A.M.; Benedik, A.I.; Torgashov, R.A.; Torgashov, G.B.; Sinitsyn, N.I. Planar Microstrip Slow-Wave Structure for Low-Voltage V-Band Traveling-Wave Tube With a Sheet Electron Beam. *IEEE Electron. Device Lett.* **2018**, *39*, 757–760. [CrossRef]
10. Torgashov, R.A.; Ryskin, N.M.; Rozhnev, A.G.; Serdobintsev, A.A.; Pavlov, A.M.; Galushka, V.V.; Bakhteev, I.S.; Molchanov, S.Y. Theoretical and Experimental Study of a Compact Planar Slow-Wave Structure on a Dielectric Substrate for the W-Band Traveling-Wave Tube. *Tech. Phys.* **2020**, *65*, 660–665. [CrossRef]
11. Wen, Z.; Luo, J.; Li, Y.; Guo, W.; Zhu, M. A Concentric Arc Meander Line Slow Wave Structure Applied on Low Voltage and High Efficiency Ka-Band TWT. *IEEE Trans. Electron Devices* **2021**, *68*, 1262–1266. [CrossRef]
12. Specht, D.F. A general regression neural network. *IEEE Trans. Neural Netw.* **1991**, *2*, 568–576. [CrossRef] [PubMed]
13. LeCun, Y.; Bengio, Y.; Hinton, G. Deep learning. *Nature* **2015**, *521*, 436–444. [CrossRef]
14. Tahersima, M.H.; Kojima, K.; Koike-Akino, T.; Jha, D.; Wang, B.N.; Lin, C.W.; Parsons, K. Deep Neural Network Inverse Design of Integrated Photonic Power Splitters. *Sci. Rep.* **2019**, *9*, 1368. [CrossRef] [PubMed]
15. Lin, R.H.; Zhai, Y.F.; Xiong, C.X.; Li, X.H. Inverse design of plasmonic metasurfaces by convolutional neural network. *Opt. Lett.* **2020**, *45*, 1362–1365. [CrossRef] [PubMed]
16. Hegde, R.S. Deep Learning: A new tool for photonic nanostructure design. *Nanoscale Adv.* **2020**, *2*, 1007–1023. [CrossRef]
17. Ma, W.; Liu, Z.; Kudyshev, Z.A.; Boltasseva, A.; Cai, W.S.; Liu, Y.M. Deep learning for the design of photonic structures. *Nat. Photonics* **2021**, *15*, 77–90. [CrossRef]
18. Ma, W.; Cheng, F.; Liu, Y.M. Deep-Learning-Enabled On-Demand Design of Chiral Metamaterials. *ACS Nano* **2018**, *12*, 6326–6334. [CrossRef] [PubMed]
19. Alagappan, G.; Png, C.E. Deep learning models for effective refractive indices in silicon nitride waveguides. *J. Opt.* **2019**, *21*, 035801. [CrossRef]
20. Malkiel, I.; Mrejen, M.; Nagler, A.; Arieli, U.; Wolf, L.; Suchowski, H. Deep learning for the design of nano-photonic structures. In Proceedings of the 2018 IEEE International Conference on Computational Photography (ICCP), Pittsburgh, PA, USA, 4–6 May 2018; pp. 1–14.
21. Chen, T.; Guestrin, C. XGBoost: A Scalable Tree Boosting System. In Proceedings of the 22nd ACM SIGKDD International Conference on Knowledge Discovery and Data Mining, San Francisco, CA, USA, 13–17 August 2016; pp. 785–794.
22. Yi, D.; Ahn, J.; Ji, S.M. An Effective Optimization Method for Machine Learning Based on ADAM. *Appl. Sci.* **2020**, *10*, 1073. [CrossRef]
23. Rumelhart, D.; Hinton, G.; Williams, R. Learning representations by back-propagating errors. *Nature* **1986**, *323*, 533–536. [CrossRef]
24. Guo, G.; Wei, Y.; Zhang, M.; Travish, G.; Yue, L.; Xu, J.; Yin, H.; Huang, M.; Gong, Y.; Wang, W. Novel Folded Frame Slow-Wave Structure for Millimeter-Wave Traveling-Wave Tube. *IEEE Trans. Electron. Devices* **2013**, *60*, 3895–3900. [CrossRef]

Article

Study of an Attenuator Supporting Meander-Line Slow Wave Structure for Ka-Band TWT

Hexin Wang [1], Shaomeng Wang [1,*], Zhanliang Wang [1], Xinyi Li [2], Tenglong He [1], Duo Xu [1], Zhaoyun Duan [1], Zhigang Lu [1], Huarong Gong [1] and Yubin Gong [1,*]

1 National Key Lab on Vacuum Electronics, University of Electronic Science and Technology of China, Chengdu 610054, China; whx427@126.com (H.W.); wangzl@uestc.edu.cn (Z.W.); faithhill@foxmail.com (T.H.); xuduo1234567@hotmail.com (D.X.); zhyduan@uestc.edu.cn (Z.D.); lzhgchnn@uestc.edu.cn (Z.L.); hrgong@uestc.edu.cn (H.G.)

2 Nanjing Sanle Electronic Group Co., Ltd., Nanjing 211800, China; leexy222@163.com

* Correspondence: wangsm@uestc.edu.cn (S.W.); ybgong@uestc.edu.cn (Y.G.); Tel.: +86-137-0900-5056 (S.W.); +86-138-0803-6055 (Y.G.)

Abstract: An attenuator supporting meander-line (ASML) slow wave structure (SWS) is proposed for a Ka-band traveling wave tube (TWT) and studied by simulations and experiments. The ASML SWS simplifies the fabrication and assembly process of traditional planar metal meander-lines (MLs) structures, by employing an attenuator to support the ML on the bottom of the enclosure rather than welding them together on the sides. To reduce the surface roughness of the molybdenum ML caused by laser cutting, the ML is coated by a thin copper film by magnetron sputtering. The measured S_{11} of the ML is below −20 dB and S_{21} varies around −8 dB to −12 dB without the attenuator, while below −40 dB with the attenuator. Particle-in-cell (PIC) simulation results show that with a 4.4-kV, 200-mA sheet electron beam, a maximum output power of 126 W is obtained at 38 GHz, corresponding to a gain of 24.1 dB and an electronic efficiency of 14.3%, respectively.

Keywords: meander-line; surface roughness; S-parameters; slow wave structure; traveling wave tube

Citation: Wang, H.; Wang, S.; Wang, Z.; Li, X.; He, T.; Xu, D.; Duan, Z.; Lu, Z.; Gong, H.; Gong, Y. Study of an Attenuator Supporting Meander-Line Slow Wave Structure for Ka-Band TWT. *Electronics* **2021**, *10*, 2372. https://doi.org/10.3390/electronics10192372

Academic Editor: Yahya M. Meziani

Received: 16 August 2021
Accepted: 26 September 2021
Published: 28 September 2021

1. Introduction

With the large-scale application of 5G communication networks and the rapid development of 6G, the demands for high-power, high-efficiency millimeter wave sources are increasing rapidly [1–5] in recent years. Compared with the widely used solid-state power amplifiers (SSPAs), the traveling wave tube has inherent advantages at a millimeter wave frequency, such as generating high-power electromagnetic waves easily and with high efficiency and an excellent heat dissipation ability [6–9].

However, the traditional TWT usually has disadvantages, such as large size and weight, high operation voltage and so on, which make it difficult to meet the requirements of the communication industry for miniaturization and mass production. As a result, the planar slow wave structure (SWS) has attracted wide attention over the last decade. One kind of self-winding helix quasi-planar SWS is explored and fabricated by using MEMS technology for potential mass production [10]. The planar SWS features as almost two-dimensional rather than three-dimensional. The typical planar SWS, formed by periodically bending a thin metal line on the dielectric substrate, is a kind of microstrip meander line. This process can be seen as flattening the traditional helix SWS into a flat structure. In the category of planar SWSs, many different bending forms, such as V-shaped [11], U-shaped [12], angular log-periodic [13], coplanar [14], etc., have been proposed successively.

The microstrip SWSs have the advantages of a low operation voltage and easy fabrication by using semiconductor technology for mass production, yet there are some problems needing to be solved before applying them in practice to TWTs. For example, the metal layer is usually too thin (less than 10 μm) to withstand the heat generated by high energy electrons bombardments [15], and the accumulating charges on the dielectric substrate tend to damage the SWSs [16].

Therefore, a meander-line (ML) SWS supported by side dielectric rods instead of a bottom substrate slab was proposed [17] and furtherly developed with different patterns [18,19]. The ML which was cut off from a metal sheet by laser is much thicker (~200 μm) than those fabricated by magnetron sputtering. As a result, it not only inherits the advantages of the low voltage and small size of the microstrip type SWSs, but also solves the problems mentioned above. Based on this, the staggered dual ML SWS supported by side dielectric rods is proposed in [20], in which the simulation results show it can generate 283 W at 75 GHz. In addition, the diamond bottom supported ML SWSs are also studied and fabricated in the Ka-band [21] and X-band [22], in which the measurement of transmission characteristics shows a comparable agreement with the simulations. In [23], a new fabrication method to make the side dielectric rods supporting the ML SWS, is explored by using mechanical roll bending of the copper strip. However, there was a problem found on the ML SWS, that the feasibility was difficult to guarantee when welding the metal meander line and dielectric rods. The assembly process is still complex.

Therefore, in order to simplify the assembly process, a novel attenuator supporting meander-line (ASML) SWS is presented in this paper. Its main feature is using the dielectric attenuator block to support the ML at the bottom, instead of welding them together on the side, which makes the connection between the ML and dielectric rods much easier. Instead of using traditional helix rods with an attenuator, the dielectric attenuator block simplifies the assembly process and retains the function to absorb the backward oscillation at the same time. In addition, by reducing the distance between the bottom metal shield and the ML, the operation voltage is further reduced. According to the simulation results, with the electron beam voltage of 4.4 kV, an output power of 126 W can be obtained from the ASML SWS TWT, with a maximum gain and electronic efficiency of 24.1 dB and 14.3%, respectively.

In addition, the fabrication processing and the measurement of the surface roughness and reflection-transmission characteristics of this structure are discussed. By using a laser confocal microscope to measure the surface roughness, it is verified that the surface roughness has been improved after copper coating on the ML's original material of molybdenum (Mo). The measured S-parameters characteristics of the SWS shows during the designed band of 36–39 GHz, the S_{11} is below -20 dB and the S_{21} of the ML varies around -8 to -12 dB and the S_{21} of the attenuator is below -40 dB. This new structure inherits the characteristics of the ML, such as high output power, planarization and low working voltage, further reducing the difficulty of processing and assembly which makes it easier to manufacture and mass produce.

The rest of the paper is organized as follows: in Section 2, the structure model and parameters are presented, as well as the high frequency characteristics of the ASML SWS. In Section 3, the input and output structures are designed and fabrication issues are discussed. In addition, the surface roughness and S-parameters measurement results are presented and analyzed. In Section 4, the results of beam-wave interaction obtained from particle-in-cell (PIC) simulation are shown to predict the potential performance of a TWT based on the ASML SWS. At the end, Section 5 gives the conclusion of the article.

2. Structure Model and High Frequency Characteristics of the ASML SWS

In order to study the dispersion characteristics of the ASML, a single period model is established. Figure 1 shows the perspective view of the one-period structure with labelled dimensional parameters and the cross-sectional view showing the relative positions of the meander-line, electron beam and metal shield, respectively. The metal meander-line is in a U shape with right angle corners. The thickness of the meander-lines is t, the width is a, and the length of the long arms is d. The distance between the adjacent straight lines is j, so the length of a single period in z direction is $p = 2 \times (a + j)$.

The height of the metal shield is *ay*, the distance from the bottom metal shield is *dy*. When *dy* is small, it can be regarded as loading the ridge on the bottom layer, which reduces the working voltage.

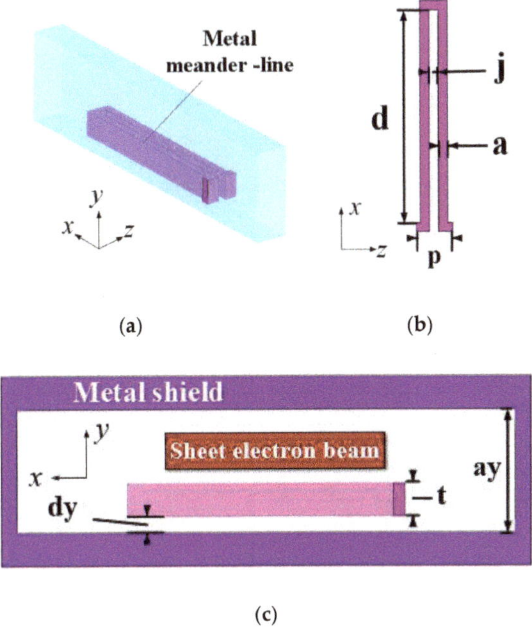

Figure 1. The different view of one period of the ASML SWS in the (**a**) 3-D, (**b**) xz plane and (**c**) xy plane.

Different from the PDU-MML SWS [12], this kind of SWS is supported from the bottom by the attenuator located around the middle of the SWS instead of welding the line and dielectric rods, as shown in Figure 2. Similar to the fabrication of the conventional helix TWTs, the attenuator can be obtained by evaporating a carbon film on the boron nitride (BN) block easily. The dimensional parameters' values after optimization are provided in Table 1.

Table 1. The critical dimensional parameters of the ASML SWS.

Parameters	Value (mm)	Parameters	Value (mm)
d	1.6	*ay*	0.756
a	0.07	*dy*	0.1
j	0.07	*d1*	0.7
p	0.28	*d2*	0.2
t	0.2	*d3*	0.4

The dispersion curves of the ASML SWS are calculated by using the eigenmode solver of the commercial simulation software CST STUDIO SUITE. For the dimension values in Table 1, the effects of *dy* on the dispersion curves are investigated and plotted in Figure 3a. It can be seen that the curves become flat when *dy* reduces, which means the phase velocity of the electromagnetic wave is becoming smaller. Finally, the *dy* = 0.1 mm is used in the later TWT design. The Figure 3b indicates the interaction impedance is above 40 Ohms from 10–40 GHz.

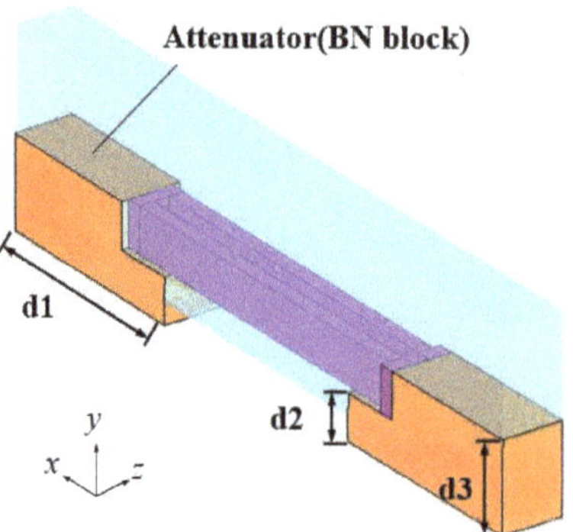

Figure 2. Geometry of the attenuator part in 3-D view.

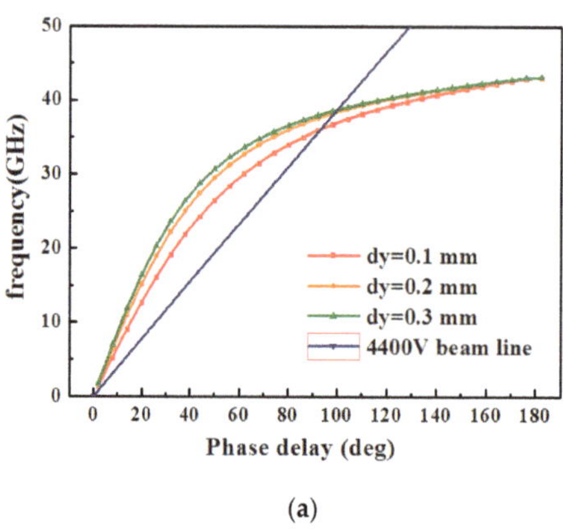

(**a**)

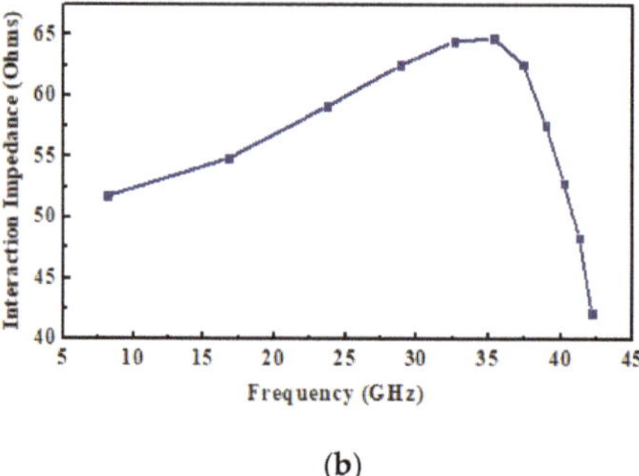

(**b**)

Figure 3. (**a**) The dispersion curves versus frequency for different values of dy and (**b**) the interaction impedance.

3. Fabrication and Measurement Issues

3.1. Fabrication and Assembly

Figure 4a gives the assembly sketch of the ASML SWS, and as can be seen, the metal enclosures (pink parts) with transition stepped ridge waveguides and flanges are fabricated together with the same processing in order to reduce the assembly error as much as possible. The upper and lower metal enclosures will be fixed by the clamps (blue parts). As for the metal meander-line, the fabrication process contains two steps. The first step is through the picosecond laser cutting molybdenum sheet to fabricate the required pattern of the ML.

The second step is covering the whole ML with a thin copper film (~3 μm) by magnetron sputtering, which can improve the surface roughness and the conductivity of the ML.

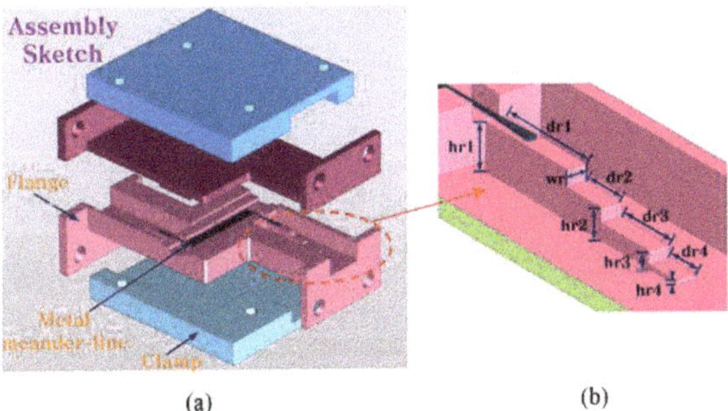

(a) (b)

Figure 4. (**a**) The assembly sketch of ASML SWS with clamp, (**b**) the input-output couplers of the stepped ridge waveguide transition.

In order to connect the ASML SWS to a standard WR-28 rectangular waveguide (3.556 mm × 7.112 mm), the stepped ridge waveguide to strip line transition that could gradually transform the TE_{10} waveguide mode to quasi-TEM mode has been proposed as input-output couplers for beam-wave interaction. Figure 4b shows the half part of the SWS and the detailed transition structure with main dimensional parameters for designing the suitable mode converter. The values of the transition structure are listed in Table 2.

Table 2. The main dimensions of the staggered stepped ridge waveguide transition.

Parameters	Values (mm)	Parameters	Values (mm)
dr1	6	*hr1*	2.3
dr2	2.55	*hr2*	1.5
dr3	3.5	*hr3*	0.8
dr4	2	*hr4*	0.2
wr	1.1	-	-

Figure 5 shows the fabricated SWS, in which the ML has already connected with the ridge of the input-output couplers by using spot welding. There are two sets of SWSs fabricated for measuring the main transmission loss of the ML with and without the attenuator, respectively. One is the ML with normal BN block (white block in Figure 5a). The other is the ML with the attenuator block that is fabricated by evaporating carbon on a BN block, which is shown in Figure 5b.

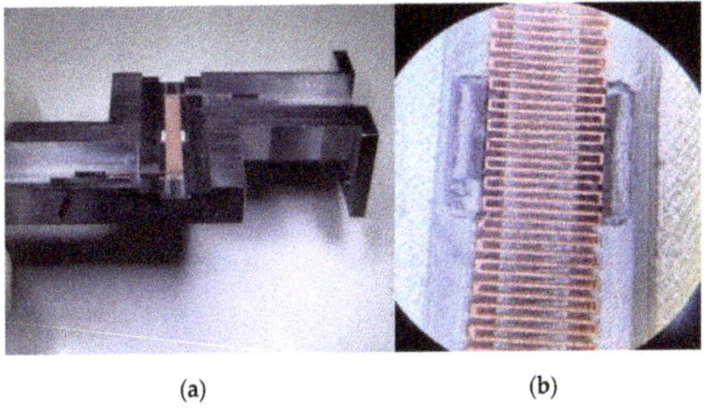

(a) (b)

Figure 5. The two sets of the fabricated SWSs, (**a**) the ML with normal BN block, (**b**) ML with attenuator block.

3.2. The Surface Roughness Measurement

To demonstrate the improvement on the surface roughness after magnetron sputtering of the copper film on the Mo meander-line, the values of roughness are measured by using a laser confocal microscope (Olympus LEXT 5000). Because the ML is cut by the laser from the top, the side surfaces are much rougher than the top surface. Figure 6a shows the side surface of the Mo ML, in which the laser cutting strip marks are obvious. The surface roughness of three random sample points varies from 410 to 670 nm. Figure 6b shows the side surface of the Mo ML after the copper film sputtering. It can be seen that the laser cutting marks are not obvious, and the values vary from 220 to 360 nm. As for the top surface of the ML, the surface roughness of the Mo ML varies from 90 to 170 nm, and the Mo ML with copper film varies from 70 to 120 nm.

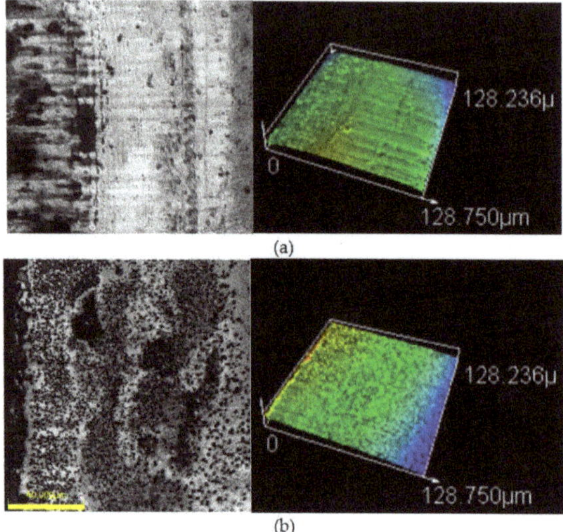

Figure 6. Test image of surface roughness by Olympus LEXT 5000 at the side surface of (**a**) Mo ML and (**b**) Mo ML with copper film.

According to the calculation equation of effective conductivity (σ_{ef}), Equation (1) [24],

$$\sigma_{ef} = \frac{\sigma}{\left(1 + \frac{2}{\pi}\arctan\left(1.4 \times \left(\frac{R_S}{\delta}\right)^2\right)\right)^2} \tag{1}$$

It can be seen that the effective conductivity can be improved by two aspects. One is increasing the material conductivity, as the conductivity of copper (5.8×10^7 S/m) is much better than that of Mo (2×10^7 S/m); the other one is to reduce the surface roughness (R_S) that is improved by coating copper as demonstrated above. The δ is skin depth, which is 339 nm at 38 GHz. According to the measured surface roughness, the effective conductivity of the ML with copper film is calculated to be around 3×10^7 S/m.

3.3. S-Parameters Measurements

To study the reflection-transmission characteristics of the 54-period ML with BN block and attenuator block supporting, respectively, two sets of SWSs are fabricated (Figure 5) and measured (Figure 7). For comparison, the simulations are also conducted by using CST STUDIO SUITE, in which the conductivity of the ML varies from 2×10^7 S/m to 5×10^7 S/m. The relative permittivity and loss tangent of the BN material are set as $\varepsilon_r = 4$ and $\tan\delta = 0.0005$, respectively.

The comparisons of the measured and simulated S_{11} and S_{21} results are shown in Figure 8a. The S_{11} results show a good agreement between the measured and simulated results, which is below −20 dB from the designed frequency band of 36 GHz to 40 GHz. As for the S_{21}, the measured result varies around −8 dB to −12 dB from 36 GHz to 39 GHz, which shows a good agreement with the simulation of conductivity of 3×10^7 S/m. In addition, this result also approximately meets the result calculated by using the surface roughness measurement results.

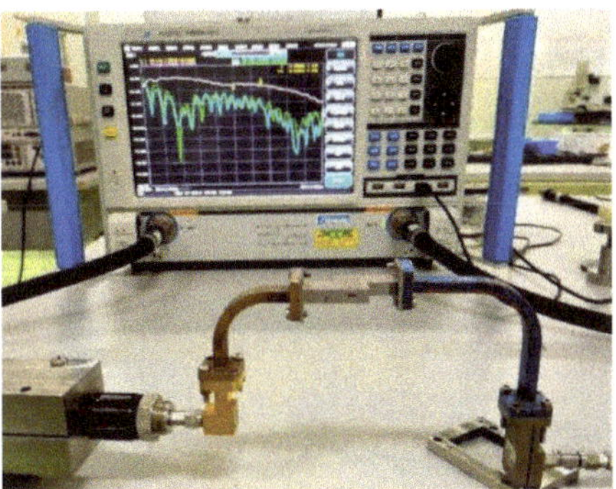

Figure 7. The measurement of S-parameters by using a vector network analyzer.

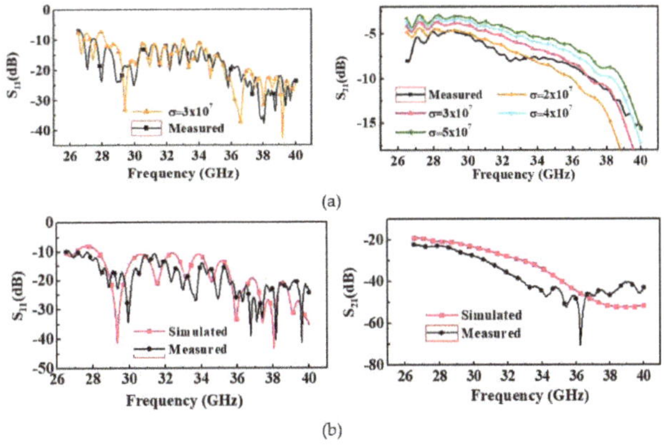

Figure 8. The comparison of S-parameters of the ML SWS with (**a**) BN block supporting and (**b**) attenuator supporting.

As for the ML with attenuator support, the relative permittivity of the attenuator is set to $\varepsilon_r = 4$ and the loss tangent is set to $\tan\delta = 0.5$ in the simulation. According to Figure 8b, it can be seen that the S_{11} stays approximately the same between the measured and the simulated results, with or without the attenuator. With regard to the S_{21}, during the 35–39 GHz, the measured and the simulated results are both below −40 dB, which means that the attenuator absorbs the majority of the wave energy and works well.

4. "Hot" Performance of the ASML TWT

To fully study the beam-wave interaction characteristics of the ASML TWT, 3-D particle-in-cell (PIC) simulations are carried out by CST PARTICLE STUDIO. A sheet electron beam with an operation voltage and current of 4.4 kV and 200 mA is used in the simulation. The cross-sectional dimensions of the sheet-beam are 960 μm × 100 μm with a current density of 208 A/cm^2. A solenoid magnetic field of 0.4 T is used to maintain the transmission of the sheet electron beam. The conductivity of the metal line is set to $\sigma = 3 \times 10^7$ S/m for eventual ohmic losses due to surface roughness.

Figure 9 shows the time and frequency domain information of the output signal at 38 GHz, in which an input signal with an average power of 0.5 W is used. After 2 ns the output signal becomes stable with a power of 126 W, corresponding to a maximum gain of 24.1 dB and an electronic efficiency of 14.3%. In the 20 ns simulation time, as shown in Figure 9a, the output signal also stays stable, showing that no oscillation occurred.

The output spectrum in Figure 9b shows the fundamental is 60 dB higher than the second harmonic. The beam trajectories and phase-space diagram of the strongly modulated electron beam are shown in Figure 10. It indicates a strong beam-wave interaction, and the wave gets plenty of energy from the electron beam.

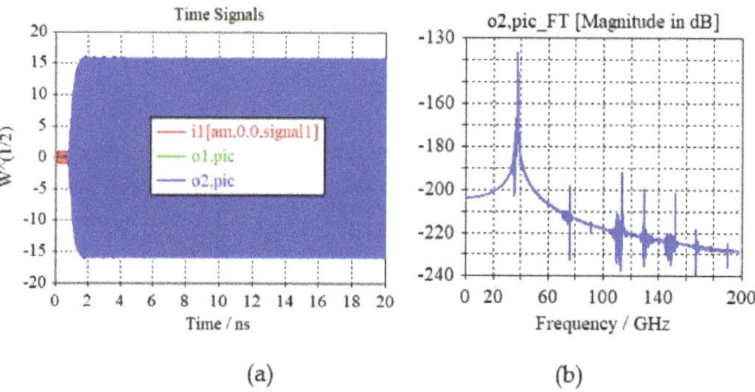

Figure 9. Output signal in (**a**) time-domain and (**b**) frequency-domain.

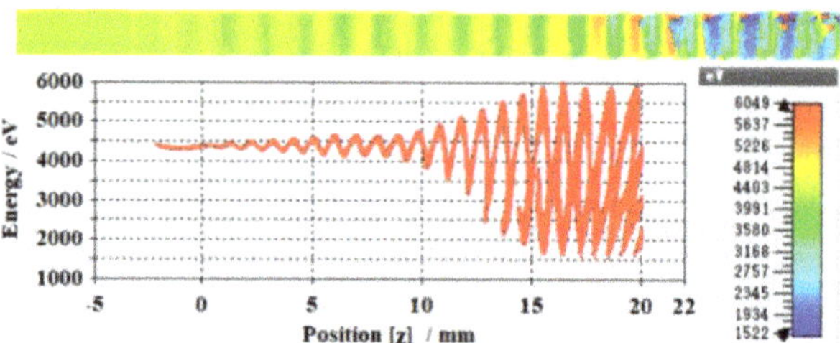

Figure 10. Electron beam bunching and phase space diagram.

Figure 11a shows the output power at 38 GHz as a function of input power with different values of beam current. When the current is 200 A, the output power increased from 5 W to 126 W, as the input power increased from 0.01 W to 0.5 W. However, in the case of the current of 0.1 A, no saturation is observed. In order to study this phenomenon, the further simulation is conducted and the results are shown in Figure 11b, which indicates that the output power at 38 GHz increased from 5 W to 185 W, as the current increased from 0.05 A to 0.25 A, when the input power is 0.5 W. Considering the beam focusing issues, a 0.2 A beam current is used for further study.

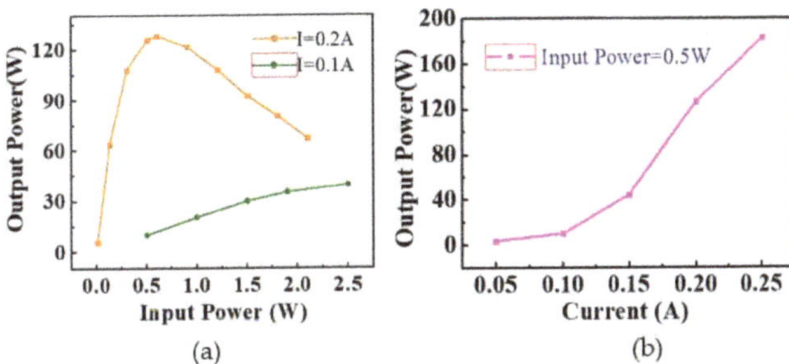

Figure 11. Variation of output power at 38 GHz with (**a**) input power and (**b**) beam current.

Figure 12 shows the variation of the output power with frequency for different beam voltages. The 3-dB hot-bandwidth for 4.4 kV is ~2.5 GHz. Moreover, thanks to the relatively wide cold bandwidth of the ASML, the center frequency can shift from 35 GHz to 40 GHz by changing the beam voltage.

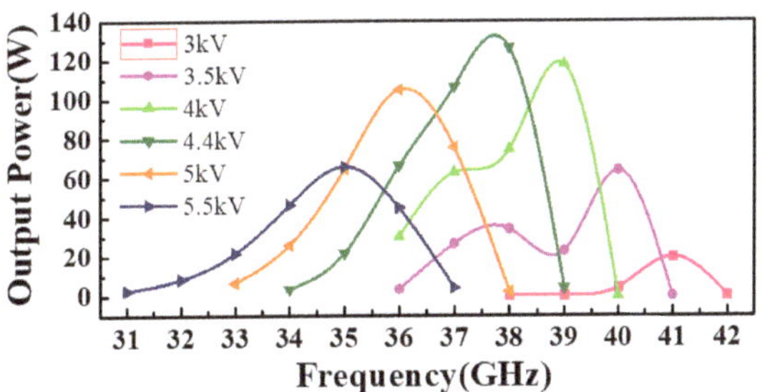

Figure 12. Output power vs. frequency for different beam voltages.

5. Conclusions

The ASML SWS for Ka-band planar TWTs has been proposed and investigated. The structure model and high frequency characteristics of the ASML SWS have been presented. The stepped ridge waveguide to strip-line couplers, which has been designed for connecting the standard Ka-band waveguide, is described. The fabrication of the ML and assembly of the ASML SWS are discussed. To improve the surface roughness of the ML, a thin copper film is coated on the original material Mo by using magnetron sputtering. The surface roughness measurement also verifies this improvement quantitatively. The S-parameters characteristics of the ASML SWS are also studied experimentally. The measured results show that during the designed band of 36–39 GHz, the S_{11} is below −20 dB and the S_{21} of the ML varies around −8 dB to −12 dB and the S_{21} of the SWS with the attenuator is below −40 dB. Furthermore, the "hot" characteristics of the ASML SWS TWT are studied by simulation. The PIC simulation results show the maximum output power of 126 W at 38 GHz is obtained with using a 4.4 kV and 0.2A beam. The maximum gain and electronic efficiency are 24.1 dB and 14.3%, respectively.

Author Contributions: Conceptualization, H.W.; validation, S.W., Z.W.; data curation, X.L., T.H., D.X.; writing—original draft preparation, H.W.; writing—review and editing, Z.D., Z.L., H.G. and Y.G.; supervision, Y.G.; project administration, Y.G. All authors have read and agreed to the published version of the manuscript.

Funding: This work was supported by National Natural Science Foundation of China (NSFC) (Grant Nos. 61921002, 61988102); Science and Technology on High Power Microwave Laboratory Fund (Grant No 6142605180201).

Conflicts of Interest: The authors declare no conflict of interest.

References

1. Jiang, W.; Han, B.; Habibi, M.A.; Schotten, H.D. The Road Towards 6G: A Comprehensive Survey. *IEEE Open J. Commun. Soc.* **2021**, *2*, 334–366. [CrossRef]
2. Hong, W.; Jiang, Z.H.; Yu, C.; Hou, D.; Wang, H.; Guo, C.; Hu, Y.; Kuai, L.; Yu, Y.; Jiang, Z.; et al. The Role of Millimeter-Wave Technologies in 5G/6G Wireless Communications. *IEEE J. Microw.* **2021**, *1*, 101–122. [CrossRef]
3. Liu, Y.; Yuan, X.; Xiong, Z.; Kang, J.; Wang, X.; Niyato, D. Federated learning for 6G communications: Challenges, methods, and future directions. *China Commun.* **2020**, *17*, 9. [CrossRef]
4. Chowdhury, M.Z.; Shahjalal, M.; Ahmed, S.; Jang, Y.M. 6G Wireless Communication Systems: Applications, Requirements, Technologies, Challenges, and Research Directions. *IEEE Open J. Commun. Soc.* **2020**, *1*, 957–975. [CrossRef]
5. Dhillon, S.S.; Vitiello, M.S.; Linfield, E.H.; Davies, A.; Hoffmann, M.; Booske, J.; Paoloni, C.; Gensch, M.; Weightman, P.; Williams, G.P.; et al. The 2017 terahertz science and technology roadmap. *J. Phys. D Appl. Phys.* **2017**, *50*, 043001. [CrossRef]
6. Basu, R.; Rao, J.M.; Le, T.; Letizia, R.; Paoloni, C. Development of a D-Band Traveling Wave Tube for High Data-Rate Wireless Links. *IEEE Trans. Electron Devices* **2021**, *68*, 9. [CrossRef]
7. André, F.; Racamier, J.C.; Zimmermann, R.; Le, Q.T.; Krozer, V.; Ulisse, G.; Minenna, D.F.G.; Letizia, R.; Paoloni, C. Technology, Assembly, and Test of a W -Band Traveling Wave Tube for New 5G High-Capacity Networks. *IEEE Trans. Electron Devices* **2020**, *67*, 7. [CrossRef]
8. Ulisse, G.; Krozer, V. W -Band Traveling Wave Tube Amplifier Based on Planar Slow Wave Structure. *IEEE Electron Device Lett.* **2016**, *38*, 126–129. [CrossRef]
9. Paoloni, C.; Gamzina, D.; Letizia, R.; Zheng, Y.; Luhmann, N.C., Jr. Millimeter wave traveling wave tubes for the 21st Century. *J. Electromagn. Waves Appl.* **2020**, *35*, 567–603. [CrossRef]
10. Prakash, D.J.; Dwyer, M.M.; Argudo, M.M.; Debasu, M.L.; Dibaji, H.; Lagally, M.G.; Van Der Weide, D.W.; Cavallo, F. Self-Winding Helices as Slow-Wave Structures for Sub-Millimeter Traveling-Wave Tubes. *ACS Nano* **2020**, *15*, 1229–1239. [CrossRef] [PubMed]
11. Shen, F.; Wei, Y.; Yin, H.; Gong, Y.; Xu, X.; Wang, S.; Wang, W.; Feng, J. A Novel V-Shaped Microstrip Meander-Line Slow-Wave Structure for W-band MMPM. *IEEE Trans. Plasma Sci.* **2011**, *40*, 463–469. [CrossRef]
12. Shen, F.; Wei, Y.; Xu, X.; Liu, Y.; Huang, M.; Tang, T.; Gong, Y. U-shaped microstrip meander-line slow-wave structure for Ka-band traveling-wave tube. In Proceedings of the 2012 International Conference on Microwave and Millimeter Wave Technology (ICMMT), Shenzhen, China, 5–8 May 2012. [CrossRef]
13. Wang, S.; Gong, Y.; Hou, Y.; Wang, Z.; Wei, Y.; Duan, Z.; Cai, J. Study of a Log-Periodic Slow Wave Structure for Ka-band Radial Sheet Beam Traveling Wave Tube. *IEEE Trans. Plasma Sci.* **2013**, *41*, 2277–2282. [CrossRef]
14. Zhao, C.; Aditya, S.; Wang, S. A Novel Coplanar Slow-Wave Structure for Millimeter-Wave BWO Applications. *IEEE Trans. Electron Devices* **2021**, *68*, 1924–1929. [CrossRef]
15. Ryskin, N.M.; Torgashov, R.A.; Starodubov, A.V.; Rozhnev, A.G.; Serdobintsev, A.A.; Pavlov, A.M.; Galushka, V.V.; Bessonov, D.A.; Ulisse, G.; Krozer, V. Development of microfabricated planar slow-wave structures on dielectric substrates for miniaturized millimeter-band traveling-wave tubes. *J. Vac. Sci. Technol. B* **2021**, *39*, 013204. [CrossRef]
16. Zhao, C.; Aditya, S.; Chua, C. A Microfabricated Planar Helix Slow-Wave Structure to Avoid Dielectric Charging in TWTs. *IEEE Trans. Electron Devices* **2015**, *62*, 1342–1348. [CrossRef]
17. Wang, H.; Wang, Z.; Li, X.; He, T.; Xu, D.; Gong, H.; Tang, T.; Duan, Z.; Wei, Y.; Gong, Y. Study of a miniaturized dual-beam TWT with planar dielectric-rods-support uniform metallic meander line. *Phys. Plasmas* **2018**, *25*, 06. [CrossRef]
18. Dong, Y.; Chen, Z.; Li, X.; Wang, H.; Wang, Z.; Wang, S.; Lu, Z.; Gong, H.; Duan, Z.; Feng, J.; et al. Ka-band dual sheet beam traveling wave tube using supported planar ring-bar slow wave structure. *J. Electromagn. Waves Appl.* **2020**, *34*, 1–15. [CrossRef]
19. Wen, Z.; Luo, J.; Li, Y.; Guo, W.; Zhu, M. A Concentric Arc Meander Line Slow Wave Structure Applied on Low Voltage and High Efficiency Ka-Band TWT. *IEEE Trans. Electron Devices* **2021**, *68*, 1262–1266. [CrossRef]
20. Wang, H.; Wang, S.; Wang, Z.; Li, X.; Xu, D.; Duan, Z.; Lu, Z.; Gong, H.; Aditya, S.; Gong, Y. Dielectric-Supported Staggered Dual Meander-Line Slow Wave Structure for an E-Band TWT. *IEEE Trans. Electron Devices* **2020**, *68*, 369–375. [CrossRef]
21. Wang, Z.; Su, L.; Duan, Z.; Hu, Q.; Gong, H.; Lu, Z.; Feng, J.; Li, S.; Gong, Y. Investigation on a Ka Band Diamond-Supported Meander-Line SWS. *J. Infrared Millim. Terahertz Waves* **2020**, *41*, 1460–1468. [CrossRef]
22. Wang, Z.; Du, F.; Li, S.; Hu, Q.; Duan, Z.; Gong, H.; Gong, Y.; Feng, J. Study on an X-Band Sheet Beam Meander-Line SWS. *IEEE Trans. Plasma Sci. Devices* **2020**, *48*, 12. [CrossRef]
23. Starodubov, A.; Atkin, V.; Torgashov, R.; Navrotsky, I.A.; Ryskin, N. On the technological approach to microfabrication of a meander-line slow-wave structure for millimeter-band traveling-wave tubes with multiple sheet electron beams. In Proceedings of the Saratov Fall Meeting 2020: Laser Physics, Photonic Technologies, and Molecular Modeling, Saratov, Russia, 29 September–2 October 2020.
24. Kirley, M.P.; Booske, J.H. Terahertz conductivity of coppersurfaces. *IEEE Trans. THz Sci. Technol.* **2015**, *6*, 1012–1020. [CrossRef]

MDPI

Article

A New Method to Focus SEBs Using the Periodic Magnetic Field and the Electrostatic Field

Pengcheng Yin [1], Jin Xu [1,*], Lingna Yue [1], Ruichao Yang [1], Hairong Yin [1], Guoqing Zhao [1], Guo Guo [1], Jianwei Liu [1], Wenxiang Wang [1], Yubin Gong [1], Jinjun Feng [2], Dazhi Li [3] and Yanyu Wei [1,*]

[1] National Key Laboratory of Science and Technology on Vacuum Electronics, School of Electronic Science and Engineering, University of Electronic Science and Technology of China, Chengdu 610054, China; 201811022514@std.uestc.edu.cn (P.Y.); lnyue@uestc.edu.cn (L.Y.); ruichaoyang@foxmail.com (R.Y.); hryin@uestc.edu.cn (H.Y.); zhaogq@uestc.edu.cn (G.Z.); guoguo@uestc.edu.cn (G.G.); jianwei@uestc.edu.cn (J.L.); wxwang@uestc.edu.cn (W.W.); ybgong@uestc.edu.cn (Y.G.)

[2] National Key Laboratory of Science and Technology on Vacuum Electronics, Beijing Vacuum Electronics Research Institute, Beijing 100015, China; fengjinjun@tsinghua.org.cn

[3] Neubrex. Ltd., Kobe 6500023, Japan; dazhi_li@hotmail.com

* Correspondence: alionxj@uestc.edu.cn (J.X.); yywei@uestc.edu.cn (Y.W.)

Abstract: In this paper, a novel method, named PM-E, to focus the sheet electron beam (SEB) is proposed. This new method consists of a periodic magnetic field and an electrostatic field, which are used to control the thickness and width of the SEB, respectively. The PM-E system utilizes this electrostatic field to replace the unreliable $B_{y,off}$, which is a tiny transverse magnetic field in the PCM that confines the SEB's width. Moreover, the horizontal focusing force of the PM-E system is more uniform than that of the conventional PCM, and the transition distance of the former is shorter than that of the latter. In addition, the simulation results demonstrate the ability of the PM-E system to resist the influence of the assembly error. Furthermore, in the PM-E system, the electric field can be conveniently changed to correct the deflection of the SEB's trajectory and to improve the quality of the SEB.

Keywords: SEB; focusing system; electron optical system

Citation: Yin, P.; Xu, J.; Yue, L.; Yang, R.; Yin, H.; Zhao, G.; Guo, G.; Liu, J.; Wang, W.; Gong, Y.; et al. A New Method to Focus SEBs Using the Periodic Magnetic Field and the Electrostatic Field. *Electronics* **2021**, *10*, 2118. https://doi.org/10.3390/electronics10172118

Academic Editor: Yahya M. Meziani

Received: 8 August 2021
Accepted: 28 August 2021
Published: 31 August 2021

1. Introduction

Recently, considerable attention has been focused on vacuum electronic devices (VEDs) as a breakthrough in powerful coherent radiation source development in the terahertz wave regime of 0.1 to 1 THz [1,2], because of their high-energy conversion efficiency and large thermal power capacity. In particular, SEB has many advantages for high power vacuum electron devices, such as increasing the input DC power and reducing the beam current density proportional to the beam width [3–6]. Previous work has indicated that devices driven by SEBs could output more power. So, in many high-power millimeters wave or terahertz wave vacuum electron devices, there is a growing trend to incorporate sheet beam designs instead of solid cylindrical beams.

However, the stability of SEB transport has been recognized as one of the key technologies required for VEDs employing SEBs, resulting from the non-axisymmetric space-charge distribution shown in Figure 1. The simplest method to confine SEBs into a narrow interaction region is the uniform magnetic field. Unfortunately, the uniform magnetic field is an unstable configuration for SEBs, due to the $\vec{E} \times \vec{B}$ drift velocity shear arising from the uniform magnetic field and the transverse space-charge field. The common instabilities stemming from the drift velocity shear are deformation and diocotron instabilities. Both experiments and theories have illustrated that these instabilities may be suppressed by increasing the strength of the uniform magnetic field. However, the strength needed to achieve this stability may be too high to implement [7–11].

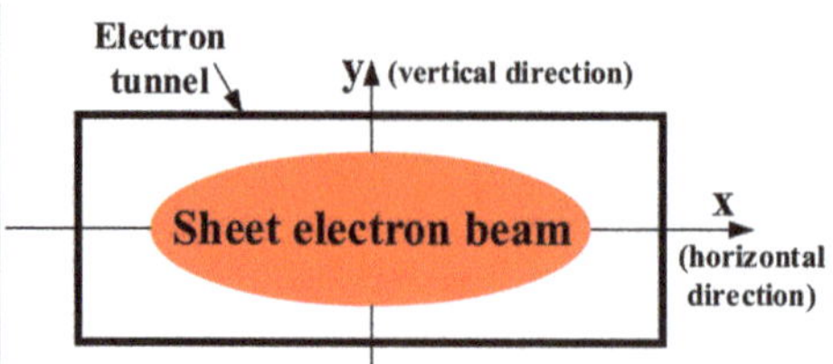

Figure 1. Cross-sectional view of SEB transportation in a tunnel.

An alternative method for focusing the SEB is the periodically cusped magnet (PCM) [12,13], which consists of periodic pole pieces and magnetic blocks. The PCM system is more suitable for focusing the SEB than the uniform magnetic system, because of its miniaturization and modest magnetic flux density. The Los Alamos National Laboratory developed a traveling wave tube (TWT) employing a 120-kV SEB focused by PCM [14]. A Ka-band TWT employing an SEB and a PCM reached a 93% transmission rate, which was designed by UESTC [15]. A Q-band SEB TWT was investigated in [16], which used a PCM system to achieve 92% beam transmission under a 30-kV beam voltage and 100-A/cm^2 current density.

Nevertheless, the sensitivity of PCM is a non-negotiable factor in experiments, resulting from a tiny transverse magnetic field. Because of the non-uniform distribution of remanence in the permanent magnetic material, it is hard to ensure that this transverse field meets the design well. Moreover, a tiny transverse magnetic field is generated by the staggered pole pieces, causing a limitation to the longitudinal magnetic field's strength.

To avoid these problems arising from the tiny transverse magnetic field, a new method, which consists of a periodic magnetic field and an electrostatic field, is presented in this paper. The electrostatic field is used to replace the tiny transverse magnetic field. For convenience, the new method is named PM-E. In addition, the PM-E system has the ability to resist assembly errors. The analysis and simulation results are presented in the following.

2. The Analysis of the PCM

The conventional PCM field with off-set pole pieces is expressed in the following form [12]:

$$\left.\begin{aligned} B_z &= B_0\cosh(2\pi y/p)\cos(2\pi/p) \\ B_{y,off} &= \frac{B_s}{\pi}[\arctan(\frac{\frac{\omega_s}{2}+x}{\left|y-\frac{b_m}{2}\right|}) - \arctan(\frac{\frac{\omega_s}{2}-x}{\left|y-\frac{b_m}{2}\right|})] \\ B_{y,pcm} &= -B_0\sinh(2\pi y/p)\sin(2\pi z/p) \end{aligned}\right\} \tag{1}$$

where $B_{y,pcm}$ and B_z are the transverse and longitudinal components of PCM fields, respectively. They are used to focus the SEB in the Y-direction. The $B_{y,off}$ represents the y component stemming from the staggered pole pieces, which is used to control the SEB in the X-direction. In fact, the distribution of $B_{y,off}$ is periodic in the Z-direction, owing to the periodicity of the pole pieces. An accurate description of the $B_{y,off}$ is shown in the following:

$$B_{y,off} = (k_a\cos(4\pi z/p) + k_a)x \tag{2}$$

where k_a denotes the coefficient of variation of $B_{y,off}$ along the Z-direction. k_b is the coefficient of variation of $B_{y,off}$ along the X-direction. Usually, because k_a is smaller than k_b, k_b plays a major role in controlling the width of the SEB. In the vicinity of the SEB, the $B_{y,off}$ can be simplified as

$$B_{y,off} = k_b x \tag{3}$$

When an electron with a longitudinal velocity v_z enters the PCM field, the interaction between the longitudinal velocity v_z and $B_{y,off}$ produces an inward force. The force can be used to balance the space-charge force, described as follows:

$$v_z B_{y,off} = E_x \tag{4}$$

Because of the ultra-high velocities of the electrons in the VEDs (the kinetic energies of the electrons are usually higher than 10 keV), $B_{y,off}$ is generally very small, only a dozen Gauss. For example, for a W-band SEB with a 0.2-A current, 19-kV voltage, and cross section of 0.2 mm × 0.8 mm, the electric field E_x arising from the space charges is 105,000 V/m at the left end of the SEB. The corresponding $B_{y,off}$ is only 13 Gs, which is much smaller than B_z (usually several thousand Gs). Considering the non-uniform distribution of remanence in the permanent magnetic material, the tiny $B_{y,off}$ tends to deviate from the ideal value, causing experimental failure. Therefore, $B_{y,off}$ is the main factor for the sensitivity of PCM.

3. PM-E

To avoid these problems stemming from $B_{y,off}$, the new method, named PM-E, which uses the electrostatic field to replace the unreliable $B_{y,off}$, is proposed. The electrostatic field is produced by the potential difference between the electron tunnel and two charged wires placed into the tunnel. The model of the PM-E system is shown in Figure 2.

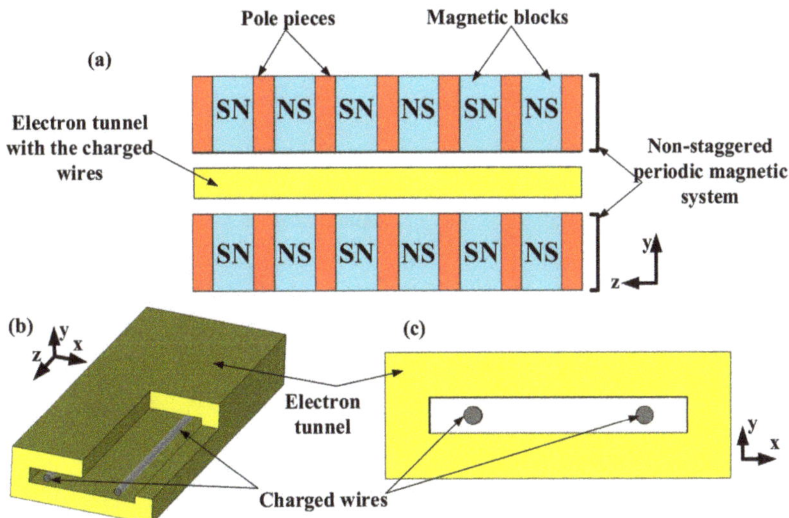

Figure 2. (**a**) The sketch of the PM-E system. (**b**,**c**) The model of the electron tunnel with two charged wires.

To resist the space-charge force in the X-direction, the potential of these two wires was slightly lower than that of the electron tunnel, as shown in Figure 3. Ordinarily, the potential difference between the charged wires and the electron tunnel would be less than 100 V. This potential difference produces an electric field opposite to the space-charge field in the X-direction, contributing to preventing the defocusing of the SEB in the X-direction.

Because $B_{y,off}$ is replaced by the electric field, the sensitivity stemming from $B_{y,off}$ is eliminated. Moreover, the staggering of the pole pieces of the periodic magnetic system is canceled. This means that the limitation of the periodic magnetic system's width is removed, which assists in increasing the peak value of the periodic magnetic field.

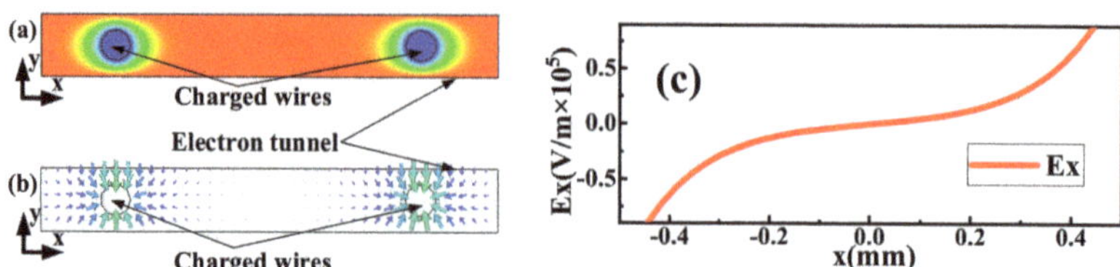

Figure 3. (**a**) The potential distribution of the electron tunnel on the X-Y plane. (**b**) The electric field distribution of the electron tunnel on the X-Y plane. (**c**) The distribution of E_x near the SEB.

According to Equation (2), $B_{y,off}$ is variable in the z-direction, while the transverse electric field provided by the PM-E system is uniform in the z-direction. In addition, the magnetic system is far from the electron channel, so the transition distance of $B_{y,off}$ is long. In contrast, the transition distance of the horizontal electrostatic focusing field is very short, due to the close distance between the two wires. The normalized horizontal focusing force of the two systems at the (−0.4 mm, 0 mm) coordinate is displayed in Figure 4. This figure indicates that the normalized horizontal focusing force of the PM-E system is more uniform and changes rapidly, assisting in providing a high-quality focusing effect.

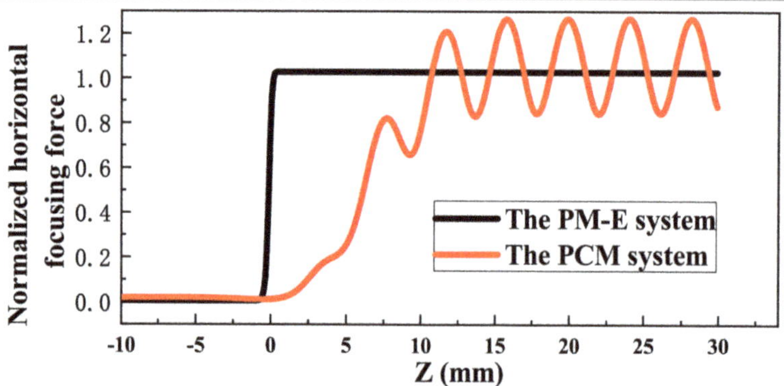

Figure 4. The normalized horizontal focusing forces of PCM and PM-E at the (−0.4 mm, 0 mm) co-ordinate.

4. Simulation

To verify the focusing effect of the PM-E system, a W-band SEB with a 0.2-A current and a 19-kV voltage was selected. The emission surface of the SEB was a 0.8 mm × 0.2 mm ellipse. The length of the tunnel was 50 mm. The dimension of the tunnel's cross section was 2 mm × 0.4 mm. The distance between the two wires with a 0.1-mm radius was 1.5 mm. The voltages of the tunnel and wires were 0 V and −60 V, respectively. The period of the magnetic system was 6 mm, and the peak value of the magnetic field was 0.15 Tesla.

The CST particle tracking solver [17] analyzed the design of the PM-E system. Figure 5 depicts the trajectory of the SEB under the influence of the field created by the PM-E system. The simulation results show that SEB could be stably transported in the PM-E field. Moreover, PM-E achieved focusing SEB in the X-direction, without $B_{y,off}$.

In addition, the PM-E system could eliminate the influence caused by assembly error, by adjusting the voltage of the wires. For example, in the above electron optical system, the left wire shifted to the right by 0.24 mm (30% of the SEB's width), as shown in Figure 6a. According to Figure 6b, under the influence of the assembly error, the trajectory of the SEB blended to the right, and the some electrons collided with the right wire.

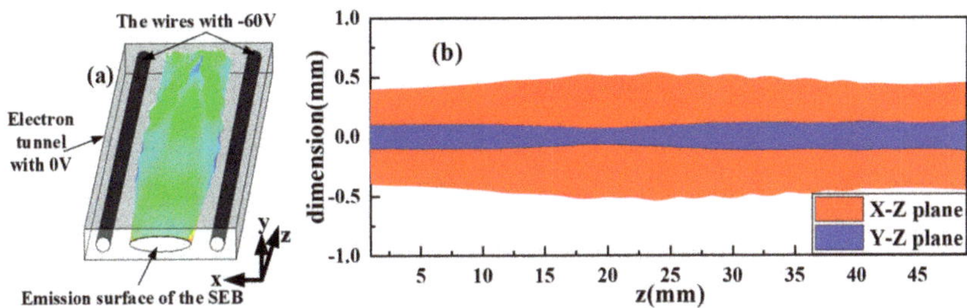

Figure 5. (**a**) 3D trajectory of the SEB analyzed by CST. (**b**) The beam trajectory on the Y−Z plane and the X−Z plane.

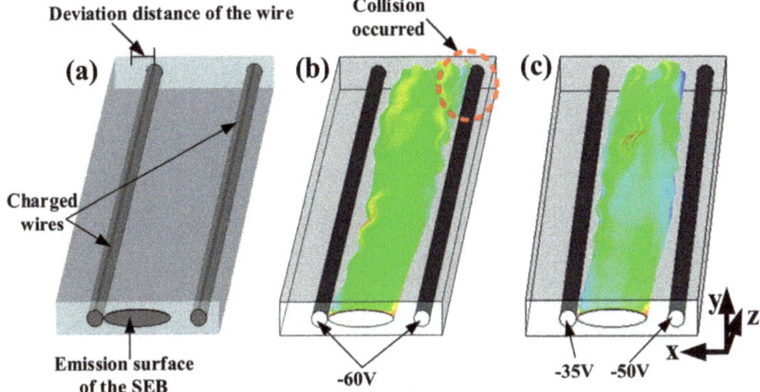

Figure 6. (**a**) The model with the assembly error. (**b**) The beam trajectory without voltage correction. (**c**) The beam trajectory with voltage correction.

In order to avoid collisions, the voltages of the two wires were set to −35 V and −50 V, respectively. As shown in Figure 6c, the direction of SEB motion was gradually corrected under the influence of the modified electric field.

Additionally, the closer the electrons were to the wire, the greater the horizontal focusing force on the electrons. Therefore, the PM-E system had a certain self-adaptive capability. In particular, when the initial position of the SEB shifted in the X-direction, the trajectory of the SEB was still confined between the two wires. For instance, Figure 7 displays that the emission surface of the SEB shifted left by 0.2 mm (25% of the width of the SEB). As can be seen in Figure 6b, the trajectory of the SEB returned to the center of the tunnel under the influence of the electrostatic field. Of course, in order to keep the SEB away from the boundary of the tunnel, the voltages of the two wires were changed slightly, and the simulation result is shown in Figure 7c. Comparing Figure 7b,c, it can be found that the quality of the SEB can be improved effectively by modifying the wire voltage when the initial position of the SEB is moved horizontally.

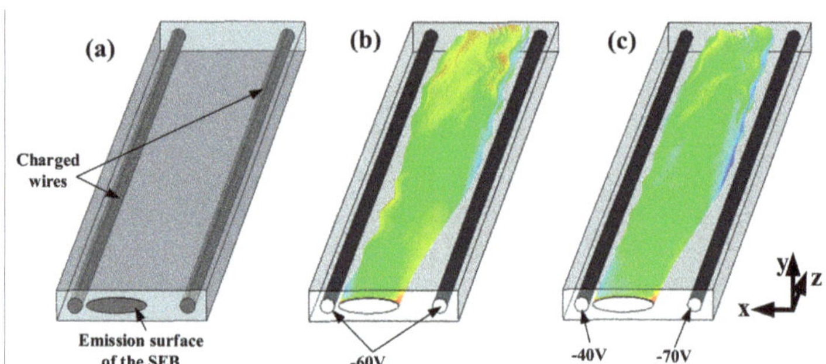

Figure 7. (**a**) A model of the emission surface moving 0.2 mm to the left. (**b**) The simulation result without changing the wire voltage. (**c**) The simulation result with changing wire voltage.

5. The Electron Optical System with the Electron Gun and PM-E System

To further verify the performance of the PM-E system, a complete electron optical system with an electron gun was designed and analyzed. The model is shown in Figure 8.

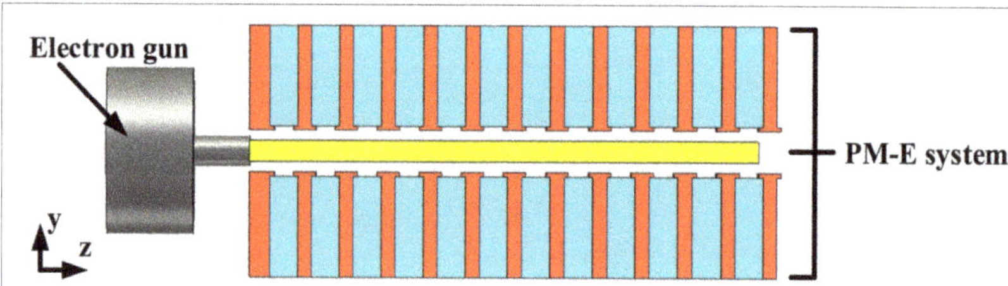

Figure 8. The model of the electron optical system with a PM-E system and electron gun.

The electron gun generated an SEB with a 19-kV voltage and 0.2-A beam current. The beam waist of the SEB was 0.2 mm × 0.8 mm with a current density of 160 A/cm^2. The emission surface of the cathode was a 1.34 mm × 1.92 mm ellipse with 9.8 A/cm^2 cathode current density. The period of the magnetic field was 8.3 mm. As shown in Figure 9, the peak value of the magnetic field was 0.22 Tesla at the axis.

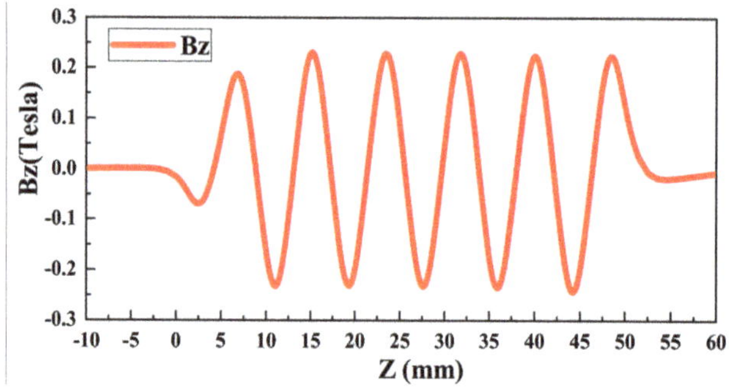

Figure 9. Distributions of the longitudinal magnetic field B_z.

The distance between the two charged wires was 1.5 mm. The voltages of the two wires with a radius of 0.1 mm were set to −60 V. The simulation result can be identified from Figure 10. And the detailed parameters used in the CST Particle Studio are listed in Table 1.

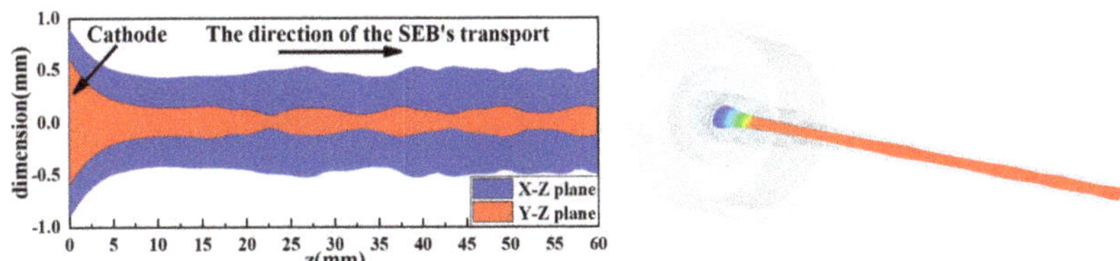

Figure 10. The simulation results of the electron optical system.

Table 1. The parameters used in CST Particle Studio.

Parameters	Values
Accuracy of the tracking solver	−50 dB
Maximum time steps	100,000
Minimum pushes per cell	5
Time step dynamic	1.2
Number of the mesh cells	13,000,000
Number of macroparticles	11,294

As can be seen in Figure 11, the cross-sectional views of SEB show that the PM-E system can maintain the laminar during transport, and the PM-E system achieves stable and efficient transport of SEB.

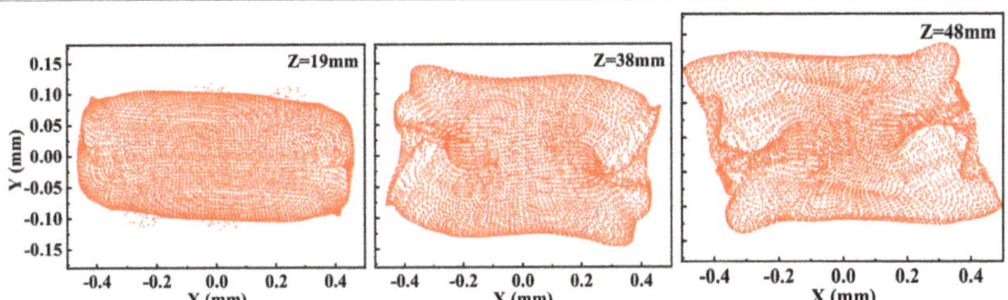

Figure 11. The cross-sectional views of the SEB in different locations.

6. Conclusions

In this paper, a novel focusing method, named PM-E, based on the electrostatic field and the periodic magnetic field, is proposed. The electric field and magnetic field control the width and thickness of SEB, respectively. The electrostatic field is produced by the potential difference between the electron tunnel and two charged wires placed into the tunnel. The periodic magnetic field is generated using a non-staggered period magnetic system. Compared with the conventional PCM system, the PM-E system utilizes an electric field instead of the unreliable $B_{y,off}$ to focus the SEB in the X-direction. The non-staggered pole pieces remove the limitation of the periodic magnetic system's width, assisting in increasing the peak value of the periodic magnetic field. Furthermore, the horizontal focusing force of the PM-E system has a uniform distribution and short transition distance. The system has a certain adaptive capacity to resist the influence of assembly errors. Furthermore, the electric field can be easily changed to correct the deflection of the trajectory and to improve the quality of the SEB. Finally, to further verify the performance of the PM-E system, a complete electron optical system with an electron gun is designed and analyzed. The simulation results demonstrate that the transmission efficiency reaches 100% under the PM-E system. As a focusing system, PM-E may have potential applications, such as traveling wave tubes and EIKs, as well as for use in other devices employing SEBs, in the future.

Author Contributions: Conceptualization, P.Y.; methodology, R.Y.; software, G.G.; validation, J.L.; formal analysis, D.L.; investigation, G.Z.; resources, R.Y.; data curation, P.Y.; writing—original draft preparation, P.Y.; writing—review and editing, P.Y.; visualization, H.Y.; supervision, L.Y.; project administration, J.X., Y.G., W.W., J.F. and Y.W.; funding acquisition, Y.W. All authors have read and agreed to the published version of the manuscript.

Funding: This work was supported in part by the National Natural Science Foundation of China (grant nos. 61988102 and 61771117).

Data Availability Statement: The data presented in this study are available on request from the corresponding author.

Conflicts of Interest: The authors declare no conflict of interest.

References

1. Komandin, G.A.; Chuchupal, S.; Lebedev, S.P.; Goncharov, Y.G.; Korolev, A.F.; Porodinkov, O.E.; Spektor, I.E.; Volkov, A.A. BWO Generators for Terahertz Dielectric Measurements. *IEEE Trans. Terahertz Sci. Technol.* **2013**, *3*, 440–444. [CrossRef]
2. Danly, B.; Petillo, J.; Qiu, J.; Levush, B. Sheet-beam Electron Gun Design for Millimeter and Sub-millimeter Wave Vacuum Electronic Sources. In Proceedings of the 2006 IEEE International Vacuum Electronics Conference Held Jointly with 2006 IEEE International Vacuum Electron Sources, Monterey, CA, USA, 25–27 April 2006; pp. 115–116.
3. Yu, D.; Wilson, P. Sheet-beam klystron RF cavities. In Proceedings of the International Conference on Particle Accelerators, Washington, DC, USA, 17–20 May 1993; Volume 4, pp. 2681–2683.
4. Pasour, J.; Nguyen, K.; Antonsen, T.; Larsen, P.; Levush, B. Solenoidal transport of low-voltage sheet beams for millimeter wave amplifiers. In Proceedings of the 2009 IEEE International Vacuum Electronics Conference, Rome, Italy, 28–30 April 2009; pp. 300–301.
5. Rusin, F.S.; Bogomolov, G.D. Orotron—An electronic oscillator with an open resonator and reflecting grating. *Proc. IEEE* **1969**, *57*, 720–722. [CrossRef]
6. Carlsten, B.E. Modal analysis and gain calculations for a SEB in a ridged waveguide slow-wave structure. *Phys. Plasmas* **2002**, *9*, 5088. [CrossRef]
7. Wang, K.; Shao, W.; Tian, H.; Wang, Z.; Lu, Z.; Gong, H.; Tang, T.; Duan, Z.; Wei, Y.; Gong, Y.; et al. Uniform permanent magnetic field with hemi-ladder structure for SEB focusing. In Proceedings of the 2018 IEEE International Vacuum Electronics Conference (IVEC), Monterey, CA, USA, 24–26 April 2018; pp. 113–114.
8. Tang, X.; Sha, G.; Duan, Z.; Wang, Z.; Tang, T.; Wei, Y.; Gong, Y. Sheet electron beam formation and transport in the uni-form magnetic field. In Proceedings of the 2013 IEEE 14th International Vacuum Electronics Conference (IVEC), Paris, France, 21–23 May 2013; pp. 1–2.
9. Panda, P.C.; Srivastava, V.; Vohra, A. Stable transport of intense elliptical SEB through elliptical tunnel under uniform magnetic field. In Proceedings of the 2011 IEEE International Vacuum Electronics Conference (IVEC), Bangalore, India, 21–24 February 2011; pp. 299–300.
10. Ruan, C.; Wang, S.; Han, Y.; Li, Q.; Yang, X. Theoretical and Experimental Investigation on Intense SEB Transport with Its Diocotron Instability in a Uniform Magnetic Field. *IEEE Trans. Electron Devices* **2014**, *61*, 1643–1650. [CrossRef]
11. Panda, P.C.; Srivastava, V.; Vohra, A. Pole-Piece with Stepped Hole for Stable SEB Transport Under Uniform Magnetic Field. *IEEE Trans. Plasma Sci.* **2015**, *43*, 2621–2627. [CrossRef]
12. Wang, J.; Liu, G.; Shu, G.; Zheng, Y.; Yao, Y.; Luo, Y. The PCM focused millimeter-wave sheet beam TWT. In Proceedings of the 2017 Eighteenth International Vacuum Electronics Conference (IVEC), London, UK, 24–26 April 2017; pp. 1–3.
13. Booske, J.H.; McVey, B.D.; Antonsen, T. Stability and confinement of nonrelativistic sheet electron beams with periodic cusped magnetic focusing. *J. Appl. Phys.* **1993**, *73*, 4140–4155. [CrossRef]
14. Carlsten, B.E.; Earley, L.M.; Krawczyk, F.L.; Russell, S.J.; Humphries, S. Stable two-plane focusing for emittance-dominated sheet-beam transport. *Rev. Mod. Phys.* **2005**, *8*, 362–368. [CrossRef]
15. Shi, X.; Wang, Z.; Tang, T.; Gong, H.; Wei, Y.; Duan, Z.; Tang, X.; Wang, Y.; Feng, J.; Gong, Y. Theoretical and Experimental Research on a Novel Small Tunable PCM System in Staggered Double Vane TWT. *IEEE Trans. Electron Devices* **2015**, *62*, 4258–4264. [CrossRef]
16. Booske, J.H.; Basten, M.A.; Kumbasar, A.H.; Antonsen, T.; Bidwell, S.W.; Carmel, Y.; Destler, W.W.; Granatstein, V.L.; Radack, D.J. Periodic magnetic focusing of sheet electron beams. *Phys. Plasmas* **1994**, *1*, 1714–1720. [CrossRef]
17. CST Studio Suite Electromagnetic Field Simulation Software, Dassault Syst., Vélizy-Villacoublay, France. 2020. Available online: https://www.3ds.com/products-services/simulia/products/cst-studio-suite/ (accessed on 10 August 2021).

Article

A G-Band High Output Power and Wide Bandwidth Sheet Beam Extended Interaction Klystron Design Operating at TM_{31} with 2π Mode

Shasha Li [1], Feng Zhang [1], Cunjun Ruan [1,2,*], Yiyang Su [1] and Pengpeng Wang [1]

1 School of Electronic and Information Engineering, Beihang University, Beijing 100191, China; lishasha@buaa.edu.cn (S.L.); SY2002515@buaa.edu.cn (F.Z.); jjjbob1024@buaa.edu.cn (Y.S.); wangpengpeng@buaa.edu.cn (P.W.)

2 Beijing Key Laboratory for Microwave Sensing and Security Applications, Beihang University, Beijing 100191, China

* Correspondence: ruancunjun@buaa.edu.cn

Abstract: In this paper, we propose a high-order mode sheet beam extended interaction klystron (EIK) operating at G-band. Through the study of electric field distribution, we choose TM_{31} 2π mode as the operating mode. The eigenmode simulation shows that the resonant frequency of the modes adjacent to the operating mode is far away from the central frequency, so there is almost no mode competition in our high mode EIK. In addition, by studying the sensitivity of the related geometry parameters, we conclude that the height of the coupling cavity has a great influence on the effective characteristic impedance, and the width of the gap mainly affects the working frequency. Therefore, it is necessary to strictly control the fabrication tolerance within 2 μm. Finally, the RF circuit using six barbell multi-gap cavities is determined, with five gaps for the input cavity and idler cavities and seven gaps for the output cavity. To expand the bandwidth, the stagger tuning method is adopted. Under the conditions of a voltage of 16.5 kV, current of 0.5 A and input power of 0.2 W, the peak output power of 650 W and a 3-dB bandwidth of 700 MHz are achieved without any self-oscillation.

Keywords: extended interaction klystron (EIK); high-order mode; sheet beam; multiple gap cavity; G-band; high output power

Citation: Li, S.; Zhang, F.; Ruan, C.; Su, Y.; Wang, P. A G-Band High Output Power and Wide Bandwidth Sheet Beam Extended Interaction Klystron Design Operating at TM_{31} with 2π Mode. *Electronics* **2021**, *10*, 1948. https://doi.org/10.3390/electronics10161948

Academic Editor: Yahya M. Meziani

Received: 1 July 2021
Accepted: 8 August 2021
Published: 12 August 2021

1. Introduction

Terahertz technology has become one of the most popular technologies, and it has important applications in high-resolution imaging, medical detection, channel communication, material structure analysis and so on [1]. However, this research and development has been restricted by terahertz radiation sources, which can produce high power, high bandwidth and high efficiency and are easy to use. Therefore, the research of terahertz sources is urgent. An extended interaction klystron (EIK) is a potential terahertz source which was proposed by Chodorow and Wessel-Berg in the 1960s [2,3]. It combines the high gain of klystron with the broad bandwidth TWT, which shortens the length of the circuit [4]. Moreover, the EIK adopts a multi-gap resonator which improves the characteristic impedance and the gain bandwidth [5]. However, due to the limitation of the high frequency and precision structures, the development of EIKs is still very slow. Nowadays, with the development of science and the research of terahertz technology, EIKs are undoubtedly put forward in the direction of high power, a high frequency and a wide bandwidth with the urgent demand [6].

Many institutions in China and abroad have conducted in-depth research on EIKs. CPI (Communications & Power Industries, Canada) has been studying EIKs for many years, which in the millimeter wave band have been very mature, with their products being developed for various equipment [7,8]. In recent years, their research on EIKs in the terahertz band has also made great progress. CPI has developed EIKs with a peak output power of 52 W and operating frequency of 220 GHz [9]. In addition, the NRL (Naval Research Lab, Washington, DC, USA) first reported on the sheet beam EIK in 2007 [10]. The G-band sheet beam EIK designed by NRL has 453 W of output power under the conditions of a voltage of 16.5 kV and a current of 0.52 A [11]. The research of EIKs in China started late, and now it is mainly focused on the W-band and G-band [12,13]. The Ka-band EIK has been developed by the Institute of Electronics at the Chinese Academy of Sciences, with an average output power of 355 W and 3-dB bandwidth of 410 MHz under an operating voltage of 9 kV and current of 0.15 A [14]. Recently, they developed a W-band EIK with a maximum output power of 1.5–3 kW. Xi'an Jiaotong University designed an EIK operating in TM_{31} mode that works in the G-band with an output power of 60 W and instantaneous bandwidth of 300 MHz [15]. In short, the research of terahertz EIKs in China needs to be further improved.

In this paper, we aim to design a high-order mode structure with an output power of 500 W and 3-dB bandwidth of 600 MHz with the sheet beam EIK operating in TM_{31} mode. Compared with the traditional EIK using the fundamental mode, the high-order mode can increase the size of the structure, thus reducing the difficulty of processing and enhancing the power density. Aside from that, the current density is also reduced, so the space charge effect can be reduced and the breakdown does not occur easily. Compared with the structure in [15], our structure uses a sheet beam instead of a circular electron beam to reduce the current density and the difficulty of processing, and the simulation results show that the output power and bandwidth of our EIK are much higher. Through design, analysis and optimization, we obtain an input/output cavity and an idler cavity with a resonance frequency of about 220 GHz, uniform electric field distribution and high characteristic impedance. Considering the machining error, we study the multi-gap cavity and get the allowable range for error. In this paper, the designed sheet beam EIK consists of six multi-gap cavities of the barbell type. In addition, to expand the bandwidth, we use stagger tuning technology. To get a better output characteristic, we have conducted detailed analysis on the input power, voltage and current as well as our magnetic field. Finally, under the condition of an input power of 0.2 W, voltage of 16.5 kV, current of 0.5 A and magnetic field of 0.5 T, 650 W of output power and 700 MHz at a 3-dB bandwidth are obtained, which are the milestones for the generation of high-output power in the G-band.

2. Consideration of High-Order Mode Structure

In this paper, a high-order mode coupled cavity with periodic arrangement gaps is designed. The 3D model of the idler cavity and output cavity are shown in Figure 1a using the scheme of a sheet beam (aspect ratio is 10). Figure 1b shows the cross-sections in the xy plane, where the period between the gaps is p, the length of the gap in x direction is d in the z-axis, the width and height of the gap are wl and h, respectively, the width and height of the beam tunnel are v and u, respectively, and the width and height of the coupling cavities are wq and hu, respectively.

We chose TM_{31} mode as the operating mode, which can be compared with the traditional scheme of the fundamental mode TM_{11} in Figure 2. On the one hand, the electric field of TM_{11} mode is stronger near the upper and lower coupling cavity, and the electric field at the center gap is a little weaker, as shown in Figure 2a, which was not suitable for the sheet beam we used. In addition, with the same dimension of the TM_{31} mode cavity working at 220 GHz, the resonance frequency of the TM_{11} mode is about one third, which is far away from the G-band. However, the electric field of the TM_{31} mode is mainly distributed in the middle, and the electric field is stronger, as shown in Figure 2b. Therefore, the TM_{31} mode is more beneficial for sheet beam interaction. On the other hand, the dimensions of the structure designed using TM_{31} mode are larger, which makes it less difficult to fabricate later.

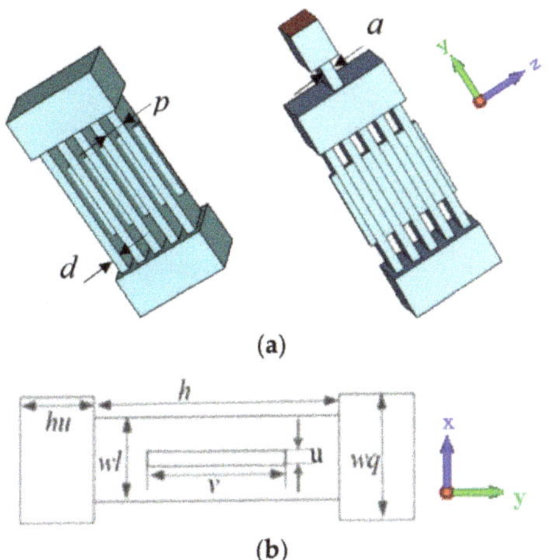

Figure 1. Schematic of the multiple gap cavity for the EIK. (**a**) A 3D model of the idler cavity and output cavity. (**b**) A cross-section in the xy plane.

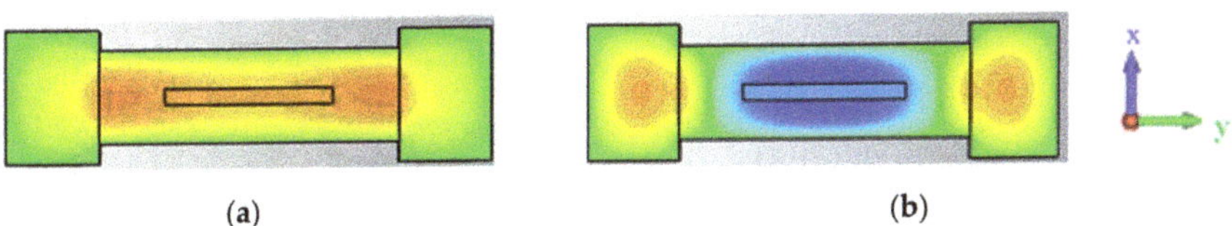

Figure 2. Electric field distribution of (**a**) fundamental TM_{11} mode and (**b**) high-order TM_{31} mode.

3. Cavity Design and Eigenmode Simulation

The distance between two adjacent gaps is p, as shown in Figure 1a, and it can be figured out from Equations (1) and (2) [16]. U_0 and f_0 are the operating voltage in kV and frequency in Hz, respectively, c is the speed of light and v_e and v_p are the electron velocity and phase velocity, respectively. Using these formulas, we can roughly calculate the initial parameters we need, and then the structure is further optimized by three-dimensional CST software [17]:

$$v_e = c\sqrt{1 - \frac{1}{(1 + U_0/511)^2}} \quad (1)$$

$$p = v_e/f_0 \quad (2)$$

As a figure of merit, the effective characteristic impedance $(R/Q)\cdot M^2$ is chosen to evaluate the beam–wave interaction ability of the EIK [18]. In the expression of $(R/Q)\cdot M^2$, R/Q and M are the characteristic impedance and coupling coefficient, respectively [19]:

$$\frac{R}{Q} = \frac{\left(\int_{-\infty}^{\infty} |E_z| dz\right)^2}{2\omega W_s} \quad (3)$$

$$M = \frac{\int_{-\infty}^{\infty} E_z e^{j\beta_e z} dz}{\int_{-\infty}^{\infty} |E_z| dz} \quad (4)$$

where E_z, W_s, ω, and β_e are the axial electric field, total stored energy, angular frequency and electronic wave number, respectively.

Through a large amount of simulation, the optimized geometry parameters of the idler cavity with five gaps are given in Table 1, which corresponds with Figure 1. Then, the electric field distribution of the five-gap idler cavity is shown in Figure 3. As can be seen from the Figure 3a, the phase difference between the adjacent gaps is 2π, and the electric field distribution accords with the TM_{31} mode. Additionally, the electric field distribution is mainly concentrated in the upper and lower coupling cavities and the middle of the gap. The electron channel is located in the region of a strong electric field in the middle of the gap, which enhances the beam–wave interaction.

Table 1. Optimized parameters of idler cavity.

Parameter	Value (μm)	Parameter	Value (μm)
d	160	wl	760
p	340	wq	1100
u	140	h	2500
v	1400	hu	760

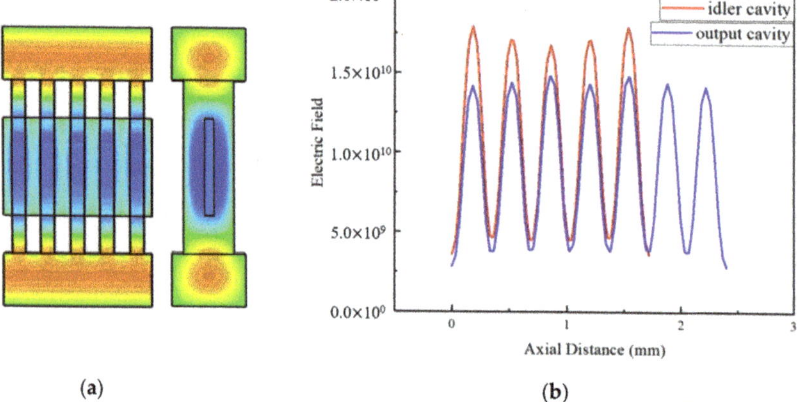

Figure 3. Electric field distribution for the 5-gap idler cavity and 7-gap output cavity. (**a**) Electric field distribution and mode pattern for 5-gap idler cavity. (**b**) Axial electric field distribution for idler and output cavities.

For the seven-gap output cavity, the dimensions and location of the coupling hole are important factors, affecting the electric field distribution and performance, especially Q_e, as shown in Figure 1a. Different widths of the coupling hole and different positions of the coupling hole were studied and, it was found that when the width was 0.2 mm and the coupling hole was located in the middle of the upper surface of the coupling cavity, the electric field distribution of the output cavity was strong and uniform, and the Q_e value was the most appropriate for the EIK. Figure 3b shows the axial electric field distribution. Whether in the idler cavity or the output cavity, the electric field should be evenly distributed. Aside from that, the electric field in the gap is stronger, while between the adjacent gaps it is weak, which provides the possibility of getting good output results. Table 2 gives the performance values of the operating mode and its adjacent modes. From the table, we can see that only $(R/Q)\cdot M^2$ of the operating mode near 220 GHz is high enough and reasonable for beam–wave interaction. Moreover, the frequency interval between the operating mode and the adjacent modes is very large, so there is almost no possibility for mode competition in the EIK.

Table 2. The performance of the operating mode and its adjacent modes.

Cavity	f (GHz)	*R/Q* (Ω)	*M*	$(R/Q)\cdot M^2$ (Ω)
5-gap idler cavity	216.76	84	0.02	0.22
	219.70	124	0.30	10.60
	223.68	97	0.07	0.48
7-gap output cavity	217.21	166	0.014	0.03
	219.30	151	0.29	10.67
	223.65	133	0.02	0.023

4. Sensitivity Analysis for the Cavity

In the mm wave and THz frequency band, devices are compact and miniaturized. Thus, small changes in dimensions may have a great impact on device performance. However, errors are inevitable in manufacturing, so an acceptable fabrication tolerance should be considered thoroughly. Figure 4 shows the influence of manufacturing error on the frequency and R/Q of an EIK with several typical geometry dimensions, (e.g., wl, wq and hu). It can be seen from Figure 4a that the change of wl has the greatest influence on the frequency. The 2-μm changes in wl will cause a frequency variation of about 300 MHz. The width and height of the coupling cavity (wq and hu) have little effect on the frequency, especially wq. Therefore, wl can be used to adjust the frequency in a large range, and hu can be used to adjust the frequency in a small range. From Figure 4b, it can be seen that the change of wl and wq has little effect on the characteristic impedance, while hu has a little greater effect on the characteristic impedance. Therefore, in the structural design, wl can be adjusted to obtain the desired resonance frequency, and it can also be appropriately increased to improve the characteristic impedance. If high characteristic impedance is required, the hu should be adjusted. To sum up, if the fabrication tolerance is controlled within 2 μm, the output characteristics of the EIK will be less affected with good performance. At present, the advanced micro machining technology can provide very high machining accuracy within 2 μm, such as DIRE, UV-LIGA, WEDM and Nano-CNC. In addition, the key technique of the dynamic tuning structure must be used for each multi-gap cavity of the EIK [20], which can definitely adjust the cavity frequency to the right working condition for the optimization of output properties.

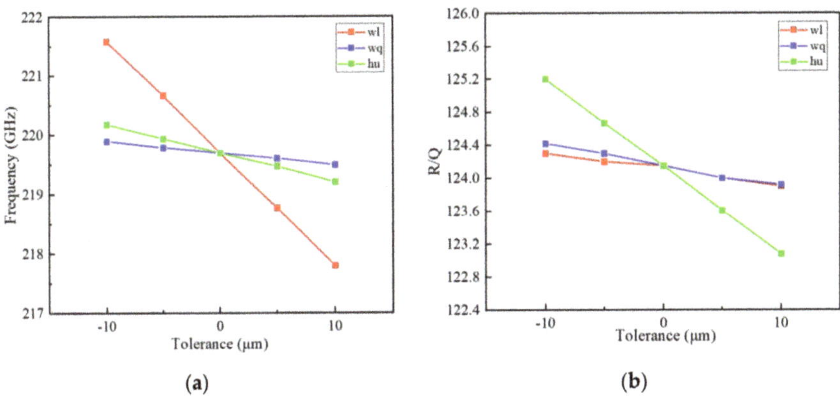

Figure 4. Influence of the fabrication tolerance of Δwq, Δwl and Δhu on the high frequency characteristics: (**a**) sensitivity of frequency and (**b**) sensitivity of R/Q.

5. Beam–Wave Interaction Simulation

Following the multi-gap cavity optimization, the beam–wave interaction system of the EIK should be analyzed thoroughly. Figure 5 shows the complete beam–wave interaction model of a six-cavity EIK we designed with CST three-dimensional software, the input cavity and the idler cavity adopted five gaps, and the output cavity adopted seven gaps. We studied the length of the drift tube between each cavity and determined the appropriate values with more analysis. In order to extend the bandwidth for our multi-gap and multi-cavity EIK, the traditional stagger tuning method was adopted in our simulation, which has been widely used for the design of a multi-cavity klystron [21–23]. The high-frequency characteristic parameters of each cavity were optimized, and they are listed in Table 3. By adjusting the value of wq, the resonant frequency of the idler cavity could be changed a little for the stagger tuning requirements, and then the best matching value of each cavity could be optimized for the beam–wave interaction analysis for our EIK.

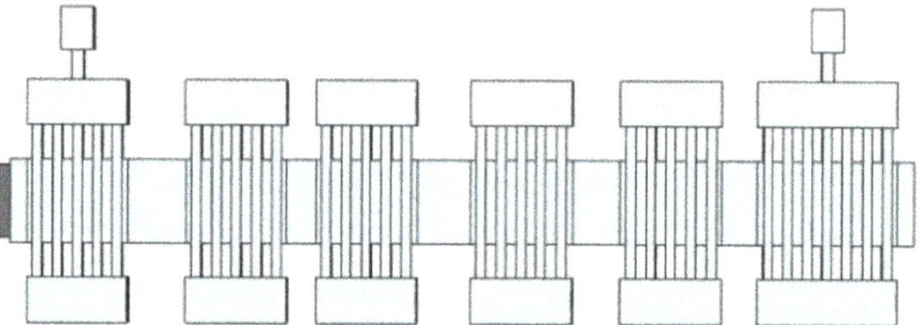

Figure 5. A 3D PIC simulation model of the complete EIK structure in CST.

Table 3. Optimized parameters of each cavity for the EIK.

N	f (GHz)	R/Q (Ω)	M	Q_e	Q_0	$(R/Q)\cdot M^2$(Ω)
1	219.64	128	0.29	200	1026	10.760
2	219.70	126	0.28	∞	1038	9.878
3	220.10	127	0.28	∞	1038	9.957
4	218.90	126	0.28	∞	1035	9.878
5	220.10	127	0.28	∞	1038	9.957
6	219.30	128	0.29	364	1026	10.760

For thorough beam–wave interaction analysis of the G-band EIK, optimization of the key parameters was performed with only one parameter change as shown in Figure 6. The relationship between the output power and 3-dB bandwidth with the input power has been given in Figure 6a. It was indicated that the output power increased first and then decreased with the increase in input power, and it reached the maximum value when the input power was 0.2 W. As the input power increased, the 3-dB bandwidth first increased and then tended to be stable. In addition, the gain decreased with the increase in the input power. Therefore, to obtain a high output power at a 3-dB bandwidth, the input power should not be too large. Therefore, we used 0.2 W as the optimized input power. With the same method, the magnitude of the magnetic field affecting the beam–wave interaction was analyzed. Figure 6b shows the relationship between the output power and magnetic field. When the magnetic field was small, as the magnetic field increased, the electron was well bound in the channel, the beam–wave interaction enhanced, and the output power would be increased. When the magnetic field was 0.5 T, the output power and bandwidth were at their maximums of 650 W and 700 MHz, respectively. Therefore, the 0.5-T magnetic field could be selected for our EIK.

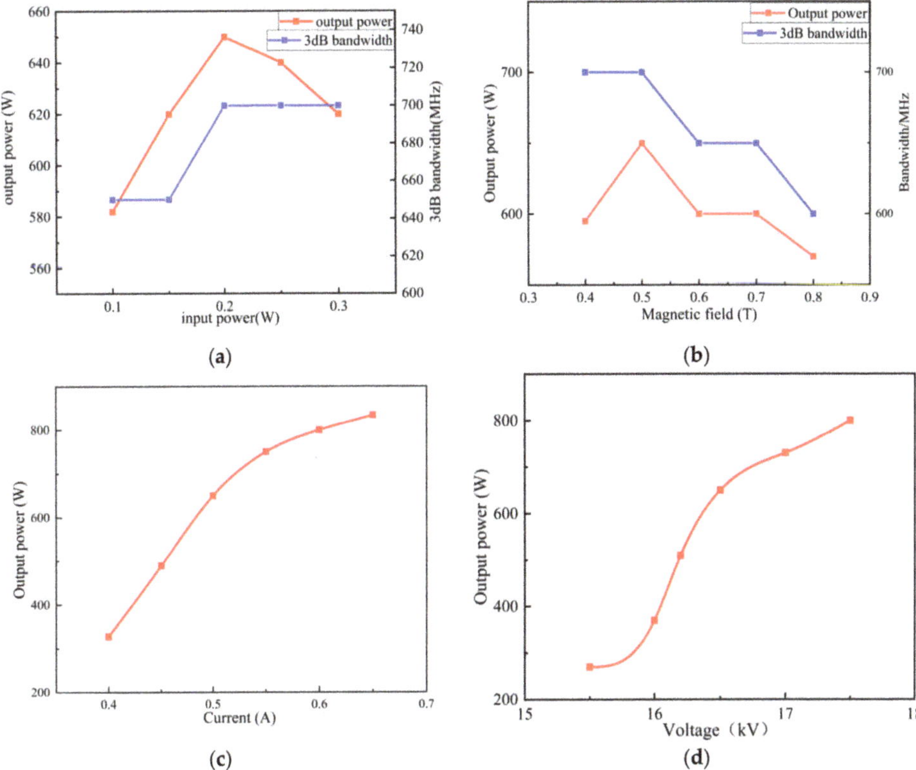

Figure 6. The variation of the output power and bandwidth with typical parameters. (**a**) The output power and 3-dB bandwidth variation with the input power, where the current (0.5 A), voltage (16.5 kV) and magnetic field (0.5 T) remain unchanged. (**b**) The output power and 3-dB bandwidth variation with the magnetic field, where the current (0.5 A), voltage (16.5 kV) and input power (0.2 W) remain unchanged. (**c**) The output power variation with the current, where the voltage (16.5 kV), magnetic field (0.5 T) and input power (0.2 W) remain unchanged. (**d**) The output power variation with the voltage, where the current (0.5 A), input power (0.2 W) and magnetic field (0.5 T) remain unchanged.

Figure 6c shows the relationship between the output power and beam current. It can be seen from the figure that as the current increased, the output power and gain increased. However, if the current was too large, although the output power would be improved, the beam focus could be more difficult, and it would also bring instability for the beam–wave interaction for the EIK. Therefore, the current of the amplifier should not be too large, and 0.5 A was selected here, with the corresponding current density being 255 A/cm^2. Figure 6d shows the relationship between the output power and voltage. With the increase in voltage, the output power also increased. If the voltage increased further, the increase in the output power would become smaller, but the voltage should not be too high. When the voltage value was small, the output was small, but the output characteristics were stable, and the time for the stable output power was short. When the voltage was too high, although the output power increased, the beam–wave interaction synchronization would be destroyed, and the output became unstable. There is a typical phenomenon in which the output power increases first, then decreases, and then it tends to be stable. As such, it will take a long time to stabilize the EIK. Thus, in this research work, the input voltage could be optimized to 16.5 kV for good performance.

After the above optimization, the structure parameters and output characteristics of the EIK were determined thoroughly. As for the results, Figure 7a shows the peak output signal at a resonant frequency of 218.9 GHz, voltage of 16.5 kV, current of 0.5 A and magnetic field of 0.5 T. It can be seen from the figure that at the beginning, the output signal gradually increased, and it soon reached the maximum value at 2.5 ns and then remained stable later. Figure 7b shows the phase space portrait of the electron energy along with the axial distance. It can be seen that in the first few cavities, due to the weak modulation effect, the energy of the electron beam changed little and fluctuated around 16 KeV, while with the enhancement of the modulation effect of the latter several cavities, the energy conversion gradually increased, and when it came to the output cavity, most electrons decelerated while a few electrons accelerated. Thus, much of the energy of the electrons would be released effectively. Figure 7c shows the bunching of the electron beam. It can be seen that the bunching effect of the electron beam from the input cavity to the output cavity was gradually enhanced, reaching its maximum in the output cavity, indicating that the electron beam had good interaction with the electromagnetic wave. Figure 7d is the frequency spectrum of the input and output signals, which shows that the amplified signal only had a peak value at 218.9 GHz, which was the same as the input frequency. In addition, the peak value of the output signal was 35 dB higher than the input signal. This shows that there was not any mode competition in the structure, and the input signal was amplified well.

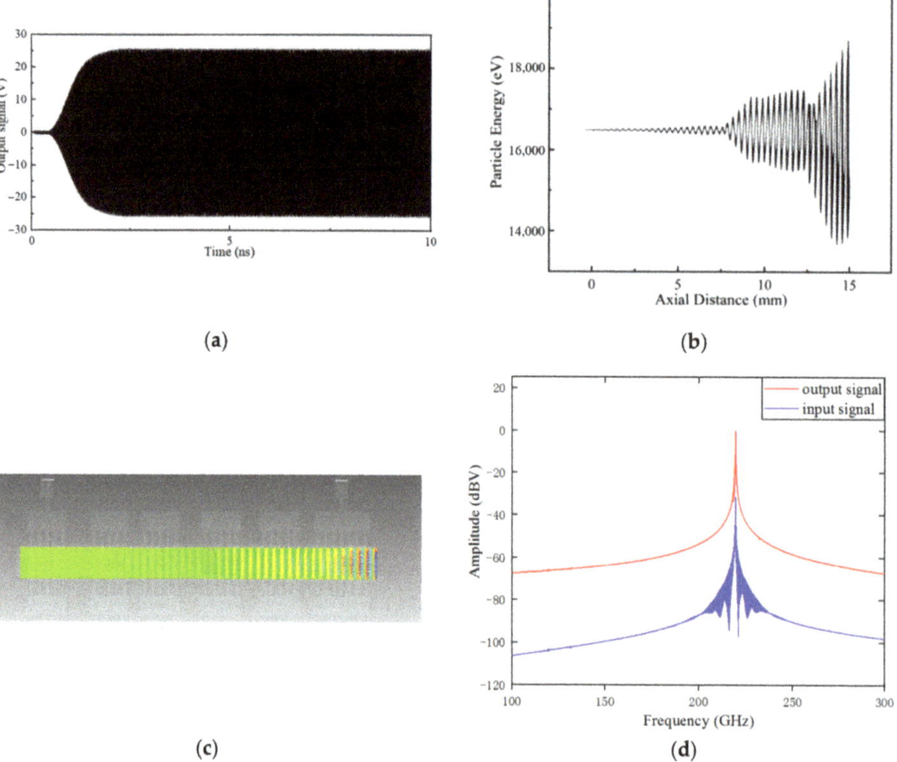

Figure 7. Output characteristics of the EIK at a frequency of 218.9 GHz: (**a**) Output signal versus time; (**b**) phase space portrait of the particle energy distribution; (**c**) beam bunching sketch; and (**d**) frequency spectrum of the input and output signals.

Figure 8 shows the relationship between the output power and frequency for our designed EIK. Under the conditions of a voltage of 16.5 kV, current of 0.5 A, focus magnetic field of 0.5 T and input current of 0.2 W, the maximum output power could reach 650 W at the frequency of 218.9 GHz. If the resonance frequency was increased or decreased, the output power would decrease gradually. Thus, the 3-dB bandwidth of 700 MHz could be achieved successfully, which was very high for a high-order mode and sheet beam EIK. Additionally, the RF efficiency of our EIK was 7.9%, and when the input power was zero, the output power of the EIK was very small, being almost zero, which shows that our structures had good suppression of self-excited oscillation, which undoubtedly made our G-band EIK have stable output performance for the beam–wave interaction and reliable output performance.

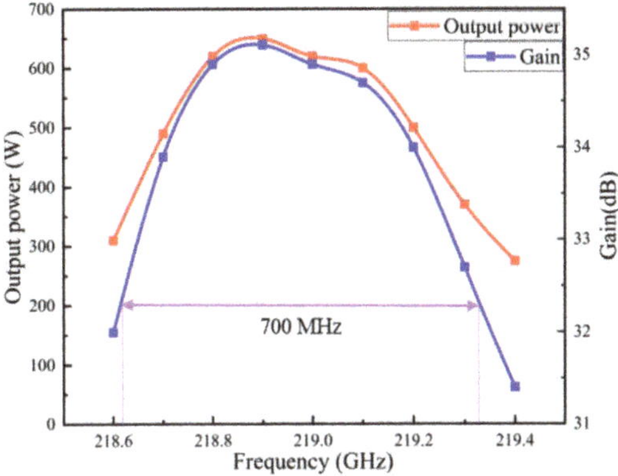

Figure 8. Output power and gain versus frequency.

The characteristic parameters of the high-order mode and fundamental mode are listed in Table 4. Compared with the circular beam EIK scheme, the sheet beam EIK with high-order mode could increase the cavity size, which reduced the difficulty of processing and also reduced the current density, so the space charge effect could be reduced. Aside from that, it could further improve the power and expand the bandwidth. Thus, our EIK with high-order mode is a good choice for high-power terahertz radiation sources [24].

Table 4. The characteristic parameters of the high-order mode EIK and fundamental mode EIK.

	f (GHz)	Cavity Size-XY (mm × mm)	Current (A)	Current Density (A/cm^2)	Voltage (kV)	Output Power (W)	Bandwidth (MHz)
Circular beam with fundamental mode	218.9	0.79 × 0.36	0.3	955	16.5	360	500
Sheet beam with high-order mode	219.2	2.50 × 0.76	0.5	255	16.5	650	700

6. Conclusions

We designed a sheet beam EIK structure working in the high-order mode TM_{31} at the G-band. This can help to increase the operation frequency with the same size of the cavity structure compared with the fundamental mode. Meanwhile, the overviewed scheme in this paper may reduce the beam current density using its big beam size and obtain a high output power and wide bandwidth. The high frequency characteristics of the multi-gap cavity were analyzed, simulated and optimized thoroughly. By adopting the stagger tuning method and six resonant cavities, we obtained good performance with a 650-W output power and 700-MHz 3-dB bandwidth with good stability. Thus, the designed physical model of our high-order mode EIK can be a good engineering choice to fabricate practical and compact radiation sources at the terahertz band in the future.

Author Contributions: Writing—original draft preparation, S.L.; writing—review and editing, S.L. and C.R.; methodology, S.L. and F.Z.; software, S.L. and F.Z.; data curation, S.L., F.Z. and P.W.; validation, F.Z. and C.R.; investigation, Y.S. and C.R.; supervision, C.R.; project administration, C.R. All authors have read and agreed to the published version of the manuscript.

Funding: This research was funded by the National Natural Science Foundation of China, grant number 61831001.

Data Availability Statement: Data is contained within the article.

Acknowledgments: The authors would like to thank J. Feng from the Vacuum Electronics National Laboratory of the Vacuum Electronics Research Institute in Beijing, China for his sincere support for the research works in this paper.

Conflicts of Interest: The authors declare no conflict of interest.

References

1. Liu, W.; Zhang, R.; Wang, Y.; Ruan, C.; Liu, P. Analysis of a two-section folded waveguide of extend interaction oscillator. In Proceedings of the 2011 IEEE International Vacuum Electronics Conference (IVEC), Bangalore, India, 21–24 February 2011; pp. 231–232. [CrossRef]
2. Chodorow, M.; Wessel-Berg, T. A high-efficiency klystron with distributed interaction. *IRE Trans. Electron Devices* **1961**, *8*, 44–55. [CrossRef]
3. Chodorow, M.; Kulke, B. An extended-interaction klystron: Efficiency and bandwidth. *IEEE Trans. Electron Devices* **1966**, *ED-13*, 439–477. [CrossRef]
4. Booske, J.H.; Dobbs, R.J.; Joye, C.D.; Kory, C.L.; Neil, G.R.; Park, G.-S.; Park, J.; Temkin, R. Vacuum Electronic High Power Terahertz Sources. *IEEE Trans. Terahertz Sci. Technol.* **2011**, *1*, 54–75. [CrossRef]
5. Zhang, C.; Ruan, C.; Wang, S.; Yang, S. High-power extended-interaction klystron with ladder-type structure. *J. Infrared Millim. Waves* **2015**, *34*, 307–313.
6. Pasour, J.; Wright, E.; Nguyen, K.T.; Balkcum, A.; Wood, F.N.; Myers, R.E.; Levush, B. Demonstration of a Multikilowatt, Solenoidally Focused Sheet Beam Amplifier at 94 GHz. *IEEE Trans. Electron Devices* **2014**, *61*, 1630–1636. [CrossRef]
7. Roitman, A.; Horoyski, P.; Hyttinen, M.; Steer, B. Wide bandwidth, high average power EIKs drive new radar concepts. In Proceedings of the Abstracts, International Vacuum Electronics Conference 2000 (Cat. No.00EX392), Monterey, CA, USA, 2–4 May 2002. [CrossRef]
8. Horoyski, P.; Berry, D.; Steer, B. A 2 GHz Bandwidth, High Power W-Band Extended Interaction Klystron. In Proceedings of the 2007 IEEE International Vacuum Electronics Conference, Kitakyushu, Japan, 15–17 May 2007; pp. 1–2. [CrossRef]
9. Hyttinen, M.; Roitman, A.; Horoyski, P.; Dobbs, R.; Sokol, E.; Berry, D.; Steer, B. A compact, high power, sub-millimeter-wave Extended Interaction Klystron. In Proceedings of the 2008 IEEE International Vacuum Electronics Conference, Monterey, CA, USA, 22–24 April 2008; p. 297. [CrossRef]
10. Nguyen, K.T.; Pershing, D.; Wright, E.L.; Pasour, J.; Calame, J.; Ludeking, L.; Rodgers, J.; Petillo, J. Sheet-Beam 90 GHz and 220 GHz Extend-Interaction-Klystron Designs. In Proceedings of the 2007 IEEE International Vacuum Electronics Conference, Kitakyushu, Japan, 15–17 May 2007; pp. 1–2. [CrossRef]
11. Nguyen, K.T.; Pasour, J.; Wright, E.L.; Pershing, D.E.; Levush, B. Design of a G-band sheet-beam Extended-Interaction Klystron. In Proceedings of the 2009 IEEE International Vacuum Electronics Conference, Rome, Italy, 28–30 April 2009; pp. 298–299. [CrossRef]
12. Xing, J.; Feng, J. Millimeter Wave Extended Interaction Device. *Vac. Electron.* **2010**, 33–37. [CrossRef]
13. Ruan, C.-J.; Wang, S.-Z.; Han, Y.; Zhang, X.-F.; Chen, S.-Y. The electron optics system and beam-wave interaction for novel W-band sheet beam klystron. *J. Infrared Millim. WAVES* **2012**, *31*, 510–516. [CrossRef]
14. Zhong, Y.; Wang, Y.; Zhang, Y. Design of Ka-band extended interaction klystron. *High Power Laser Part. Beams* **2014**, *26*. [CrossRef]
15. Wang, D.; Wang, G.; Wang, J.; Li, S.; Zeng, P.; Teng, Y. A high-order mode extended interaction klystron at 0.34 THz. *Phys. Plasmas* **2017**, *24*, 023106. [CrossRef]
16. Chen, S.; Ruan, C.; Yong, W.; Zhang, C.; Zhao, D.; Yang, X.; Wang, S. Particle-in-Cell Simulation and Optimization of Multigap Extended Output Cavity for a W-Band Sheet-Beam EIK. *IEEE Trans. Plasma Sci.* **2013**, *42*, 91–98. [CrossRef]
17. CST Corp. CST PS Tutorial. Darmstadt, Germany. Available online: http://www.cst-china.cn (accessed on 30 July 2020).
18. Shin, Y.; Park, G.; Scheitrum, G.; Caryotakis, G. Circuit analysis of Ka-band extended interaction klystron. In Proceedings of the 4th IEEE International Conference on Vacuum Electronics, Seoul, Korea, 28–30 May 2003; pp. 108–109. [CrossRef]
19. Gilmour, A.S.; Ebrary, I. *Klystrons, Traveling Wave Tubes, Magnetrons, Crossed-Field Amplifiers, and Gyrotrons*; Artech: London, UK, 2011; pp. 304–307.
20. Golde, H. A stagger-tuned five-cavity klystron with distributed interaction. *IRE Trans. Electron Devices* **1961**, *8*, 192–193. [CrossRef]
21. Symons, R.; Vaughan, R. The linear theory of the Clustered-Cavity Klystron. *IEEE Trans. Plasma Sci.* **1994**, *22*, 713–718. [CrossRef]
22. Li, R.; Ruan, C.; Zhang, H.; Haq, T.U.; He, Y.; Shan, S. Theoretical Design and Numerical Simulation of Beam-Wave Interaction for G -Band Unequal-Length Slots EIK With Rectangular Electron Beam. *IEEE Trans. Electron Devices* **2018**, *65*, 3500–3506. [CrossRef]

23. Li, R.; Ruan, C.; Li, S.; Zhang, H. G-band Rectangular Beam Extended Interaction Klystron Based on Bi-Periodic Structure. *IEEE Trans. Terahertz Sci. Technol.* **2019**, *9*, 498–504. [CrossRef]
24. Li, R.; Ruan, C.; Zhang, H. Design and optimization of G-band extended interaction klystron with high output power. *Phys. Plasmas* **2018**, *25*, 033107. [CrossRef]

 electronics

Article

Linearly Polarized High-Purity Gaussian Beam Shaping and Coupling for 330 GHz/500 MHz DNP-NMR Application

Xingchen Yang, Chaohai Du *, Ziwen Zhang, Juanfeng Zhu, Tiejun Huang and Pukun Liu *

Department of Electronics, Peking University, Beijing 100871, China; xingchenyang@pku.edu.cn (X.Y.); zhangziwen@pku.edu.cn (Z.Z.); zhujuanfeng@pku.edu.cn (J.Z.); huangtiejun_pku@pku.edu.cn (T.H.)
* Correspondence: duchaohai@pku.edu.cn (C.D.); pkliu@pku.edu.cn (P.L.)

Abstract: Terahertz waves generated by vacuum electron devices have been successfully applied in dynamic nuclear polarization enhanced nuclear magnetic resonance (DNP-NMR) technology to significantly enhance the sensitivity of high-field NMR. To reduce the magnetic field interference, the high-power terahertz wave source and the NMR spectrometer need to be separated by a few meters apart. Corrugated horns and directional couplers are key components for shaping high linearly polarized terahertz Gaussian beam and accurately coupling electromagnetic power in the transmission system. In this paper, a corrugated TE_{11}-HE_{11} mode converter and a three-port directional coupler realized by its inner cylindrical wire array are proposed for a 330 GHz/500 MHz DNP-NMR system. The output mode of the mode converter presents a characteristic of highly linear polarization, which is 98.8% at 330 GHz for subsequent low loss transmission. The designed three-port directional coupler can produce approximately −33 dB electromagnetic wave power on port 3 in the frequency range between 300–360 GHz stably, which can be used to measure the electromagnetic wave power of the transmission line in real-time. The designed mode converter and direction coupler can be installed and replaced easily in the corrugated waveguide transmission system.

Keywords: millimeter wave and terahertz; transmission line; DNP-NMR; vacuum electronics; corrugated horn; directional coupler

Citation: Yang, X.; Du, C.; Zhang, Z.; Zhu, J.; Huang, T.; Liu, P. Linearly Polarized High-Purity Gaussian Beam Shaping and Coupling for 330 GHz/500 MHz DNP-NMR Application. *Electronics* **2021**, *10*, 1508. https://doi.org/10.3390/electronics10131508

Academic Editor: Geok Ing Ng

Received: 13 May 2021
Accepted: 18 June 2021
Published: 22 June 2021

Publisher's Note: MDPI stays neutral with regard to jurisdictional claims in published maps and institutional affiliations.

1. Introduction

Nuclear magnetic resonance (NMR) technology is a spectrum measurement method, which is widely used in the fields of biomedicine and materials science [1]. However, the signal sensitivity is very low due to the small gyromagnetic ratio of nuclei such as ^{1}H and ^{13}C which hinders its further development and applications. Dynamic nuclear polarization (DNP) enhancement technology provides an effective solution to improve the sensitivity of NMR. Using DNP-NMR technology, irradiation of the THz wave on sample, due to the magnetic moment of electron is 660 times that of the ^{1}H nuclei, THz wave irradiation can enhance the electron spin polarization, electronic polarization can be easily transferred to adjacent ^{1}H nuclei, improve NMR signal sensitivity, able to reduce the original week's analysis time to a few minutes [2,3]. To obtain high-resolution spectra, modern NMR spectroscopy requires extremely high magnetic fields, thereby requiring the further development of electromagnetic (EM) wave sources used in DNP technology are correspondingly increased from millimeter to terahertz (THz) frequency bands. Vacuum electron devices (VEDs) such as gyrotrons and extended interaction klystrons (EIKs) can stably generate high-power continuous waves in the THz band [4–6]. Below 500 GHz, gyrotrons and EIKs are excellent electromagnetic sources for DNP-NMR. In order to eliminate the magnetic field interference between the THz wave source and the NMR spectrometer, they should be several meters apart. Therefore, THz transmission system proves to be a key factor for the successful application of the whole DNP-NMR system. At present, there are two kinds of transmission systems, i.e., quasi-optical transmission system and overmoded waveguide transmission system.

The free space quasi-optical beaming technique utilizes a series of lenses or mirrors to support the low loss propagation of Gaussian-like beam, but it is difficult to align stably and safely [1]. The most convenient method is using waveguides, including dielectric waveguides [7,8], corrugated metallic waveguides [9], metallic wires [10], and dielectric-lined metallic waveguides [11]. The high-order mode generated by the THz wave source is converted into the low-order circular waveguide TE_{11} mode through the mode converter. However, the ohmic loss of the TE_{11} mode is dramatically high in the THz band, and it is necessary to convert the basic circular waveguide TE_{11} mode into a low-loss mode (TEM_{00} or HE_{11}) to facilitate subsequent transmission [9,12]. A smooth profiled horn was proposed for easily manufacturing in [13] and successfully produced a Gaussian beam. Nonetheless, the converted electric field is slightly elliptical. Corrugated mode converter can efficiently transform a fundamental circular waveguide mode into the hybridHE_{11} mode, which is prevalently utilized in plasma physics [14], radio astronomy [15] and satellite communications [16], so a corrugated TE_{11}-HE_{11} mode converter is proposed to meet the requirement of DNP-NMR transmission.

Furthermore, monitoring the EM wave power stability and magnitude in real-time is necessary, because the NMR spectrometer requirements on EM wave purity and stability are strict. However, measuring the THz wave source and the EM wave power in the transmission line directly is too difficult, thereby necessitating the extraction of a part of the wave energy for the ease of measurement.

The directional coupler is widely used in the transmission line for power extraction. In a low-frequency transmission system, it is accomplished by a small coupling hole. Due to the requirements of low heat dissipation and high-power capacity in practical applications, a linear coupling holes array was proposed for the 140 GHz transmission line in Frascati-Tokamak [17]. However, when the transmitted mode is the HE_{11} hybrid mode, the power extracted by the linear holes array will be unstable. To improve the coupling stability, a quartz beam splitter was proposed to accomplish the coupling function in the 250 GHz transmission line [18]. Whereas the narrowband operation limits the performance of the quartz beam splitter for signal coupling. The metal cylindrical wire array is a potential candidate to accomplish the broadband and stability directional coupling in THz frequency. For the HE_{11} mode whose polarization direction is perpendicular to the cylindrical wire array plane, most of the waves can go across the plane and maintain their original propagation direction. For the parallel one, most of the waves are reflected. A three-port directional coupler is designed based on the cylindrical wire array and the overmoded corrugated waveguide theory in this paper. The designed three-port directional coupler can steadily extract a part of energy in a wide frequency range of 300–360 GHz.

From the DNP-NMR system shown in Figure 1, in which 500 MHz is the resonance frequency of NMR instruments, 330 GHz is the frequency of EM waves generated from the THz wave source (Figure 1 ①) and irradiated on the sample (Figure 1 ⑥). The transmission system designed in this paper is aimed to transmit 330 GHz EM waves generated by THz source to the NMR spectrometer which includes a TE_{11}-HE_{11} converter, corrugated waveguide, three-port directional coupler, and 90° miter bend. The mode converter and three-port directional coupler can be easily combined by the well-developed corrugated waveguide technology. The mode converter can generate pure HE_{11} mode, and then the HE_{11} mode will be transmitted by the low loss metal circular corrugated waveguide. A 90° miter bend can change the direction of EM wave propagation at the corner. A direction coupler extracts approximately −33 dB power in real-time to monitor the whole energy in the transmission.

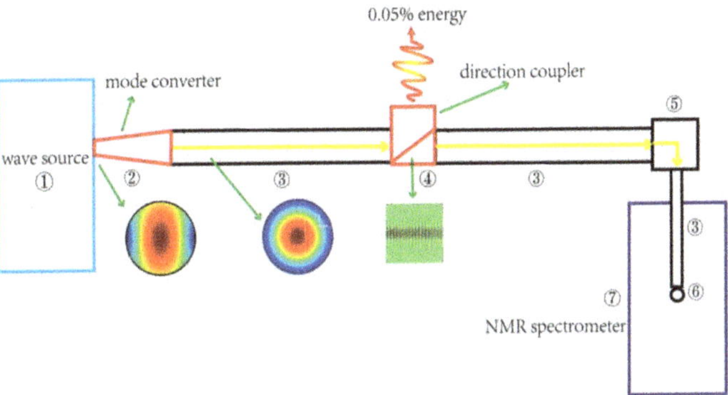

Figure 1. 330 GHz transmission line layout for DNP-NMR system. ① THz wave source, ② TE_{11}-HE_{11} mode converter, ③ circular corrugated waveguide, ④ direction coupler, ⑤ 90° miter bend, ⑥ sample, ⑦ NMR spectrometer.

2. TE_{11}-HE_{11} Corrugated Mode Converter

In our design, the input mode is TE_{11} mode, and the ideal output mode is HE_{11} mode which consists of 85% TE_{11} mode and 15% TM_{11} mode by the power ratio. As shown in Figure 2, the corrugated horn is carved with several equally spaced corrugated slots on the inner surface of the horn, and the curved contour of the horn is $\sin^2$ type. Compared with other profiles such as the linear type, $\sin^2$ profile can maintain a lower reflection over a wide frequency range [19].

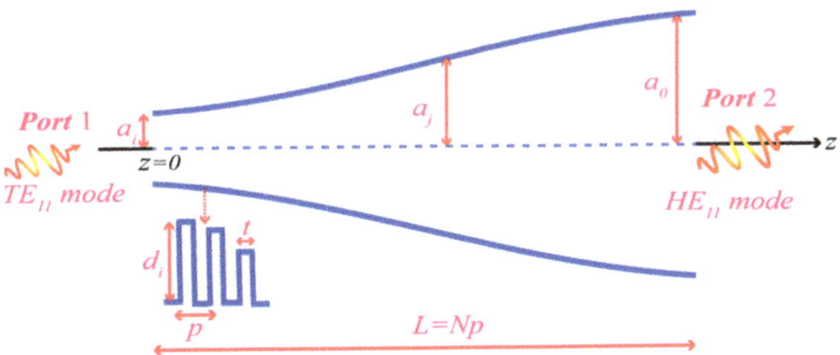

Figure 2. Schematic diagram of a TE_{11}-HE_{11} mode corrugated converter.

The purity of the output mode is controlled by several parameters, such as the radii of the input and output waveguides, the depth, width, and period of the slot. The depth of the slot decreases with the increase of the horn radius and remains constant in the end. The horn's radius $a(z)$ is expressed in (1) [14], where z is the coordinate axis, a_i is the input radius of the horn, a_o is the output radius of the horn, L is the total length of the horn, and A is a parameter that affects the similarity between the horn profile and sin^2; the smaller A is, the more the horn profile curve tends to be sin^2 shape.

$$a(z) = a_i + (a_o - a_i)\left[\frac{Az}{L} + (1-A)sin^2\left(\frac{\pi}{2}\frac{z}{L}\right)\right] \tag{1}$$

The input diameter of the mode converter is identical to the standard diameter of the fundamental circular waveguide. The input radius is around $a_i \geq 1.841c/(2\pi f_{min})$, where c is the light speed. At the same time, it should be consistent with the output window of the THz wave source. The output radius is connected to the overmoded waveguide by a tapered horn. The parameter A is around 0.2. The length L of the horn is around $2.5a_0^2/\lambda$, where λ is the wavelength at the operating frequency. There are N corrugated slot periods in the whole inner wall of the horn, so the length of L should be an integer multiple of the period, i.e., $L = Np$. The parameter p is the corrugated slot period which is around $\lambda/5$. The width of the corrugated slot is t, and $\delta = t/p$ is the ratio of the width of the corrugated slot to the period p, which is around 0.8. The slot depth along the horn is a critical parameter for horn performance [20].

The first corrugated slot depth is $d_1 = \lambda/2$, then the slot depth decreases to $\lambda/4$ through Nc cycles, with the number of Nc between 5–12. This section is called the mode conversion or impedance conversion area in which the output TM_{11} mode will be mixed with TE_{11} mode to form HE_{11} mode. After the mode conversion zone, the depth of the slot needs to introduce a correction factor so the slot depth is given by (2), where d_j is the depth of the j-th slot, which varies with the horn radius a_j, λ is the wavelength, k is wavenumber at the operating frequency, a_j is the radius at the j-th slot.

$$d_j = exp\left[\frac{1}{2.114(ka_j)^{1.134}}\right]\frac{\lambda}{4} \tag{2}$$

For the 330 GHz corrugated mode converter, the input radius $ai = 0.43$ mm, output radius $ao = 2$. Therefore, only five parameters, including N ($L = Np$), A, Nc, p, and δ ($t = \delta p$), need to be optimized. N and Nc are set in 55–75 and 5–12, respectively. The ranges of A and δ are set as 0.2–0.4 and 0.6–0.9, respectively. p is set as 0.15–0.25. The optimization is aimed at constraining the power ratio of TE_{11} mode and TM_{11} mode close to $85/15$ (≈ 5.7) to mix HE_{11} mode at the output port and reduce the reflection of the TE_{11} at the input port. Therefore, cost functions are used in the optimization and is expressed by Equation (3):

$$\mathrm{F}(N, Nc, \delta, A, p) = \frac{1}{M}\sum_{f}^{M}\frac{|S_{21}^{TE}|^2}{|S_{21}^{TM}|^2} - \frac{85\%}{15\%}, \tag{3}$$

where f values are the discrete frequency points in the desired frequency range (300–360 GHz), M is the number of the frequency sample points. $|S_{21}^{TE}|$ and $|S_{21}^{TM}|$ are the TE and TM mode magnitude at the output port, respectively. Next, we need to find out the vector $\tau = (N, Nc, \delta, A, p)$ to minimize the cost function $\mathrm{F}_{(N,Nc,\delta,A,p)}$, namely:

$$\tau = \underset{N,Nc,\delta,A,p}{\mathrm{arg}min}\ \mathrm{F}(N, Nc, \delta, A, p|a_i, a_0), \tag{4}$$

The final values of the parameters after the optimization are shown in Table 1.

Table 1. Parameters of TE_{11}-HE_{11} corrugated mode converter.

Parameter	Value
a_i	0.43 mm
a_0	2 mm
L	12.73 mm
A	0.3
N	67
Nc	6
p	0.19 mm
δ	0.7
t	0.13 mm
d_1	0.45 mm

The S_{11} parameter means the reflection of the TE_{11} mode on port 1. As shown in Figure 3, the S_{11} parameter is less than −25 dB at the range of 300–360 GHz, which shows good transmission characteristics of the horn. The S_{11} parameter can be as low as –26.6 dB at 330 GHz.

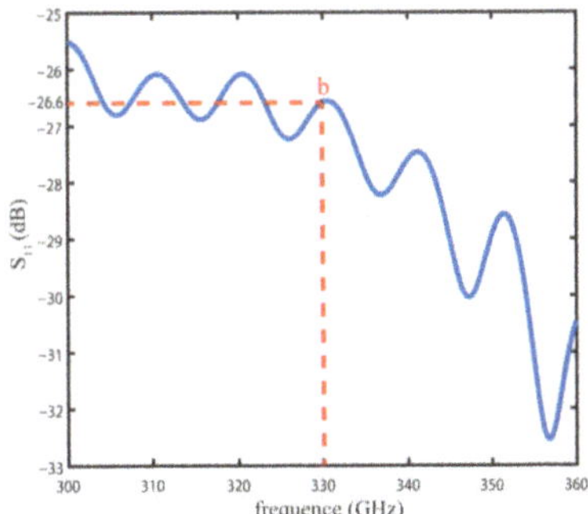

Figure 3. Simulation results of S_{11} parameter (the reflection of TE_{11} mode) ranging from 300 to 360 GHz. Point b is the S_{11} parameter whose value is −26.6 dB at the frequency of 330 GHz.

Figure 4 shows the simulated profile of the electric field at the output port, which is closed to the HE_{11} basic mode in the circular corrugated waveguide. As shown in Figure 5, the TE_{11} mode gradually shapes into HE_{11} mode while traveling through the horn. At the output port and the operation frequency 330 GHz, the power ratio of TE_{11} and TM_{11} is 6.72. The value of the electric field integral in x, y and z directions are 0.029395, 0.00023821, and 0.0032955, respectively, and the ratio of the electric field in the direction of x polarization is 98.8% at 330 GHz, which can maintain good linear polarization characteristics.

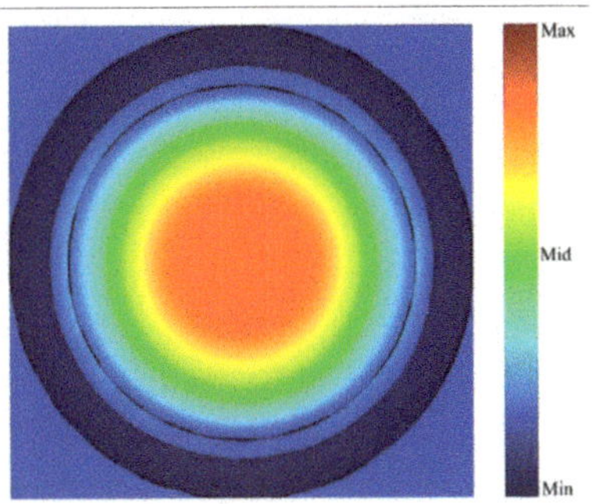

Figure 4. Electric field distribution at the output port.

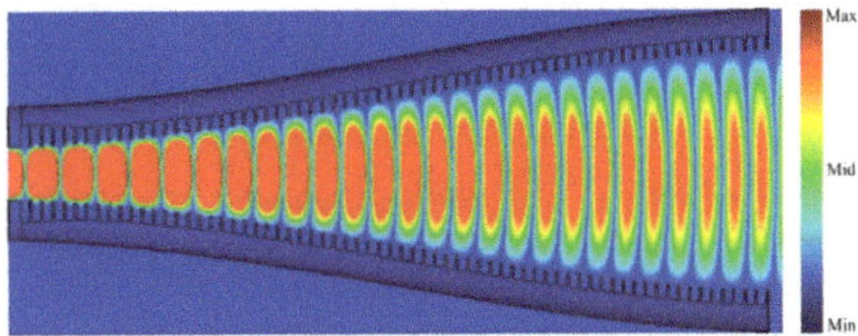

Figure 5. Cross-section electric field distribution at 330 GHz in the TE_{11}-HE_{11} mode converter.

The corrugated horn is an excellent mode converter, which can effectively convert the single mode in the metal waveguide into the HE_{11} mode. These excellent characteristics can be mainly attributed to two facts. First, the corrugated horn can realize the TM and TE modes with same phase velocity transmission so that the same phase relationship can be maintained at different frequencies, which greatly expands the working bandwidth of the corrugated horn. Second, the slot depth has a good inhibitory effect on the longitudinal current in the horn, and the electric field near the slot is sharply weakened.

3. Cylindrical Wire Array Theory and Three-Port Directional Coupler

By combining the corrugated waveguide and the cylindrical wire array theories, we designed a three-port directional coupler for the 330 GHz/500 MHz DNP-NMR transmission line. The designed three-port directional coupler is composed of a corrugated waveguide and a cylindrical wire array which is placed in the middle of the corrugated waveguide diagonal. From the 3D structure shown in Figure 6, the EM waves are fed from the port 1 split into two parts out of ports 2 and 3. All the three-port directional coupler structural parameters are shown in the sectional view by Figure 7, where a, p, t, and d are the diameter, corrugated slot period, width of the corrugated slot, and depth of the corrugated slot, respectively; N, R, and h are the cylindrical wire number, cylindrical wire diameter and period of the cylindrical wire array, respectively.

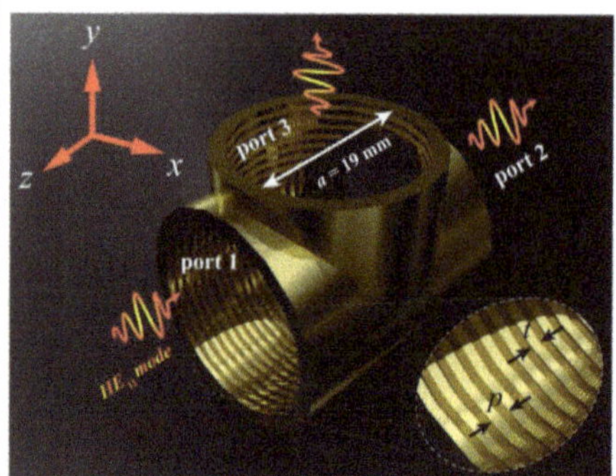

Figure 6. 3D structure of the three-port directional coupler.

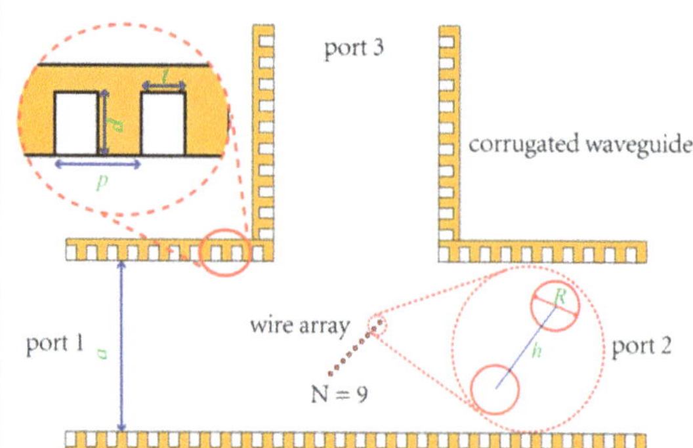

Figure 7. Sectional view of the three-port directional coupler.

For the three-port directional coupler, there are three important design indicators, i.e., (1) extraction approximately 0.05% EM wave power on port 3 stably, which can be used to measure the EM wave power of the transmission line in real-time; (2) as much EM wave as possible output at port 2, so as not to affect the power of the EM wave transmitted to the NMR spectrometer. This requires port 1 to have a very small reflection, with an ideal value of 0, and to transmit as much power to port 2 as possible, with an ideal value of 1–0.05%; (3) to make the directivity value of the three-port directional coupler larger, when the input port is port 2, as much power as possible should be output at port 1, with an ideal value of

1, and no power should be coupled out at port 3, with an ideal value of 0. To sum up, the ideally designed objective of S-matrix at 330 GHz is:

$$\begin{bmatrix} 0 & 1 & S_{13} \\ \sqrt{1-0.05\%} & 0 & S_{23} \\ \sqrt{0.05\%} & 0 & S_{33} \end{bmatrix}, \quad (5)$$

where the performance of this three-port directional coupler is independent of the parameter S_{13}, S_{23}, and S_{33}.

For the corrugated waveguide, if the following conditions are met, i.e., (1) the diameter a is much larger than λ (the wavelength at the operating frequency); (2) the corrugated slot depth and period are approximately $\lambda/4$ and $\lambda/3$, respectively; (3) $t/p > 0.5$, the field distribution in the waveguide can satisfy the equilibrium condition. Besides, the transmitted eigenmode is the HE_{11} mode with the linear polarization. The expression of the inner electric field distribution of the waveguide with a circular cross-section can be simplified as follows:

$$E_x = E_0 J_0\left(\frac{2.405\rho}{a}\right),\ E_y \approx 0, E_z \approx 0, \quad (6)$$

where J_0 is the zeroth-order Bessel function; E_x, E_y, and E_z are the components of the electric field in the x, y, and z directions; E_0 is the magnitude of the electric field; ρ is the radial coordinate. When the operating frequency is 330 GHz, a, p, t, and d are 19, 0.32, 0.16, and 0.23 mm, respectively [21].

A metal cylindrical wire array was introduced to coupling the propagating HE_{11} mode to the waveguide since the cylindrical wire array has different reflection and transmission coefficients for HE_{11} mode with different polarization, which is determined by the period of the cylindrical wire array and the radius.

As shown in Figure 8, assuming that the cylindrical wire array has an infinite number of periods along the x direction, and simultaneously, the length of the metal cylinder extends indefinitely along the z-axis, the incident direction of the Gaussian beam is at an angle of φ^{in} to the x-axis. Besides, l is l-th metal cylinder. The Gaussian beam and HE_{11} mode coupling efficiency can be as high as 98% [22]. Moreover, by Fourier transform and Gaussian integration, the expression of the Gaussian beam can be expended using the superposition of a series of plane waves. The total scattering electric field of each plane wave on the cylindrical wire array can be calculated from the single-cylinder scattering electric field.

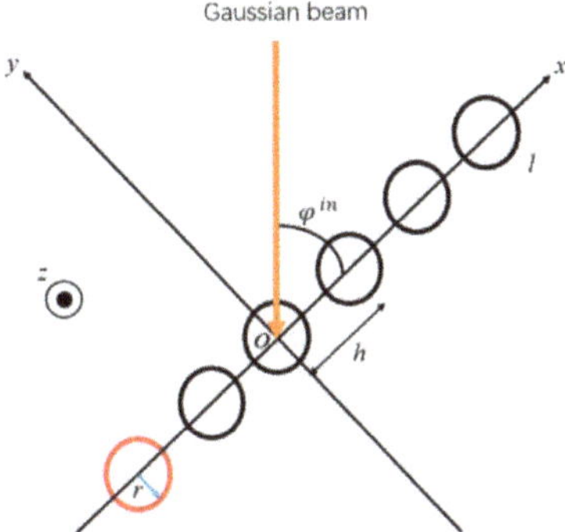

Figure 8. Schematic diagram of the periodic cylindrical wire array structure.

The transmission coefficient T and reflection coefficient R of the power can be calculated by Equations (7) and (8), where k_0 and k_y are the wave number of Gaussian beam and the component in the y direction, k_{ym} and k_{xm} are the wave number in x and y direction of m-th space harmonic, σ_l^{SG} is the harmonic coefficient of the l-th cylindrical scattering field:

$$T = \left| 1 + \frac{2}{hk_y} \sum_{l=-\infty}^{\infty} \left(-\frac{k_{ym} + ik_{xm}}{k_0} \right)^l \sigma_l^{SG} \right|^2 \tag{7}$$

$$R = \left| \frac{2}{hk_y} \sum_{l=-\infty}^{\infty} \left(\frac{k_{ym} - ik_{xm}}{k_0} \right)^l \sigma_l^{SG} \right|^2 \tag{8}$$

when the angle of EM wave direction of propagation and cylindrical wire array plane is 45°, the electric field distribution of two different linear polarization electromagnetic waves shown in Figures 9–11 show the curves of the transmission coefficient (T) and reflection coefficient (R) varying with the diameter of the cylindrical wire and the period of the wire array respectively, which are obtained by using the above theory in MATLAB when the Gaussian fundamental mode with the polarization direction is perpendicular to the wire array plane, therefore, the incident angle is 45° and the frequency is 330 GHz. It can be seen that it is highly consistent with the results obtained by two-dimensional simulation in COMSOL Multiphysics. The transmission coefficient (T) decreases with the increase of diameter and increases with the increase of period h, while the reflection coefficient is opposite to this.

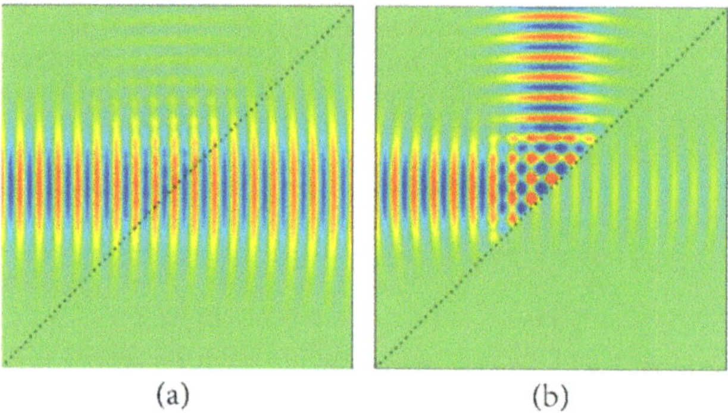

Figure 9. Electric field distribution of the incident wave when the linear polarization direction of the EM waves is perpendicular (**a**) and parallel (**b**) to the cylindrical wire array plane.

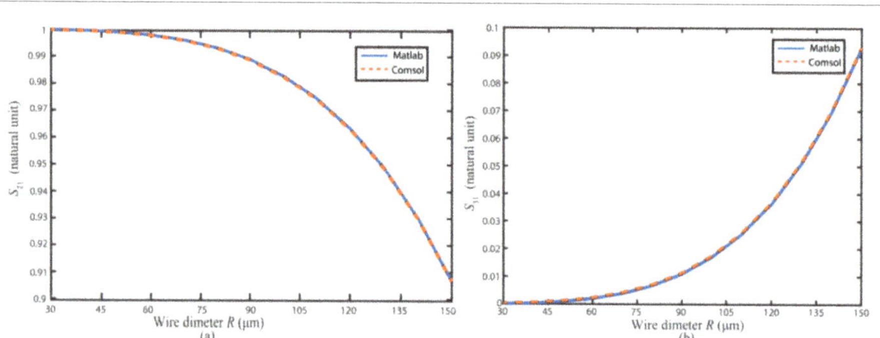

Figure 10. (**a**) Transmission (S_{21}) and (**b**) reflection (S_{31}) coefficients of Gaussian fundamental modes with the diameter of cylindrical wire R variation.

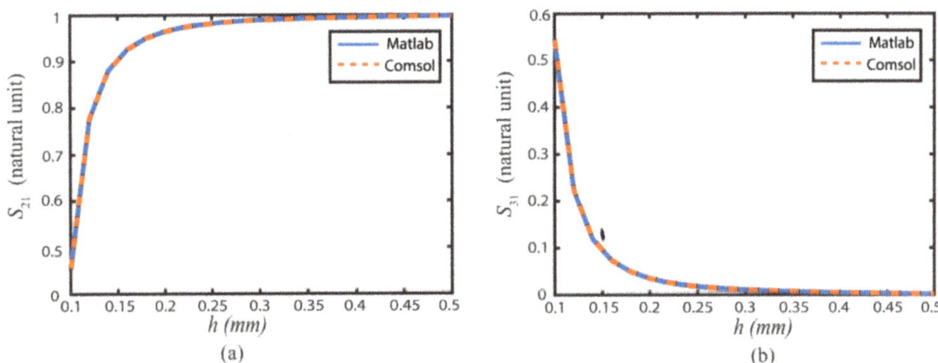

Figure 11. (**a**) Transmission (S_{21}) and (**b**) reflection (S_{31}) coefficients of Gaussian fundamental modes with the wire array period h variation.

After optimization, the period of the wire array h is 0.3 mm, and the diameter of each cylindrical wire R is 90 μm; as shown in Figure 12, in the frequency range from 300 to 360 GHz, the transmission coefficient T (S_{21}) is approximately 0.99, the reflection coefficient R (S_{31}) can be less than 0.0125. At 330 GHz, the transmission coefficient T is 0.989, and the reflection coefficient R is 0.011.

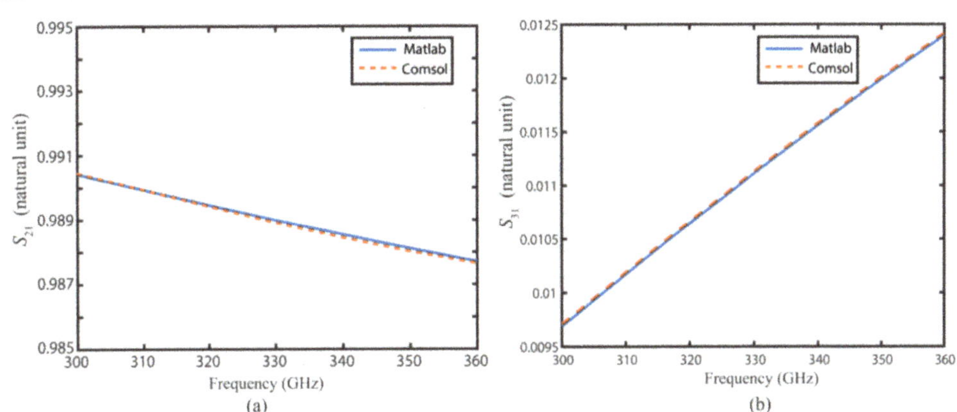

Figure 12. (**a**) Transmission coefficient T (S_{21}) and (**b**) reflection coefficient R (S_{31}) when the linear polarization direction of the EM waves is perpendicular to the cylindrical wire array plane in the frequency band ranging from 300 to 360 GHz.

As shown in Figure 13, parameter sweeping is employed with respect to the number of cylindrical wires. Our design goal is to couple a small amount of power without affecting the main transmission line. Only nine metal cylinders are required to be placed in the center of the diagonal, and the reflection coefficient R is 0.00586 at 330 GHz.

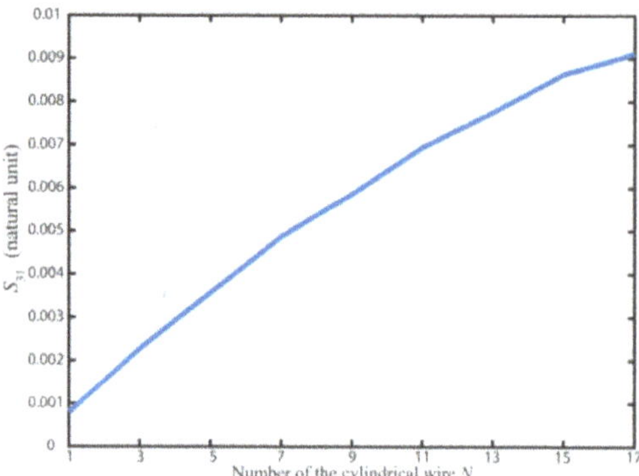

Figure 13. Reflection coefficient R (S_{31}) of Gaussian beam with a different number of cylindrical wire at 330 GHz.

We modeled and simulated the three-port directional coupler in CST Studio to verify the performance. The HE_{11} mode can be excited by mixing 85% circular waveguide TE_{11} mode and 15% TM_{11} mode together. High-purity linear polarization is required in the NMR spectrometer. For the main beam transmitted in the transmission line (HE_{11} mode wave polarization direction is perpendicular to the cylindrical wire array plane), S_{11}, S_{21}, and S_{31} parameters are shown in Figure 14; S_{12}, S_{22}, and S_{32} parameters can be obtained by converting the input port. At 330 GHz, the S_{11} parameter can reach up to −41 dB, and keep below −20 dB in the 300–360 GHz frequency range; a small part of energy output at port 3, most of the energy continues transmitting. At 330 GHz, they are −33.2 dB and −0.064, respectively, as shown in Figure 14. The coupling coefficient can remain stable in the 300–360 GHz frequency band. By setting the power value detector on port 3, the EM wave power in the transmission line can be measured in real-time. An important parameter of a directional coupler is directivity. For a three-port directional coupler, the difference in dB of the output power P3 (the power output on port 3 by port 1 input) and P3* (the power output on port 3 by port 2 input) is called directivity. The directivity of the designed three-port directional coupler is shown in Figure 15. It indicates that at 330 GHz, the directivity can reach 11.8 dB.

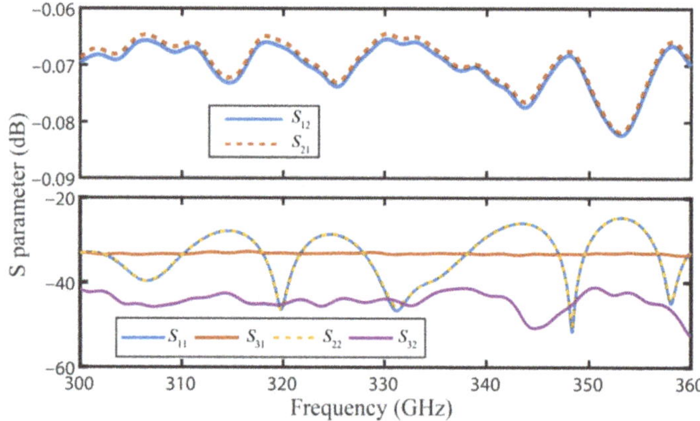

Figure 14. Simulation results of S_{11}, S_{21}, S_{31}, S_{12}, S_{22}, and S_{32} parameters from 300–360 GHz, where S_{21} is the transmission coefficient, S_{31} is the coupling coefficient.

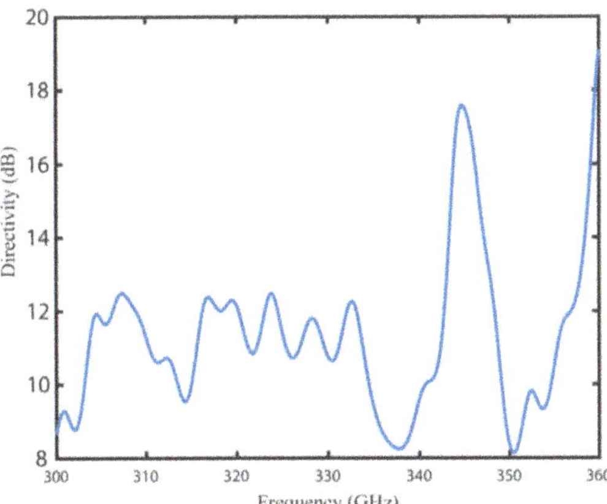

Figure 15. The directivity of the three-port directional coupler in the frequency band of 300–360 GHz.

In order to verify the integrity of the system as shown in Figure 1, a 90° bend waveguide with the same structure as the three-port directional coupler is designed. There are many design schemes for 90° miter bend, such as the principle of conversion optics and the surface plasmon polaritons [23,24]. The simplest method to realize the 90° miter bend is to directly reflect the EM wave by using the metal mirror. It has a wide working frequency band, so it is widely used in practice [25]. However, these methods do not consider the influence on the polarization direction of the propagation mode. In this paper, we simply verify a 90° miter bend based on the different reflection coefficients (R) of the electromagnetic wave with different polarization directions.

There are two differences with the three-port directional coupler, i.e., (1) as shown in Figure 9a, for the HE_{11} mode whose polarization direction is perpendicular to the cylindrical wire array plane, most of the waves can go through the wire array and maintain their original propagation direction. For the parallel one, most of the waves are reflected as shown in Figure 9b. The electric field polarization of the incident EM wave of the three-port directional coupler is perpendicular to the wire array, while the electric field polarization of the incident electromagnetic wave of the designed 90° miter bend is parallel to the wire array plane; (2) to get the maximum power output, the number of period of the wire array is 70 to cover the whole diagonal of the waveguide (a = 19 mm, h = 0.3 mm, R = 90 μm), while the three-port directional coupler N = 9.

Since most of the energy will be coupled out at the port 3, the direction of EM wave propagation can turn 90°. Therefore, this structure can achieve the function of 90° miter bend. As the result shown in Figure 16, the S_{11} parameter can be less than -50 dB at the frequency of 300–360 GHz, the S_{31} parameter can be larger than -0.15 dB. Although the S_{21} parameter is not ideal, subsequent work can be done to redesign the period and diameter of the wire array used in 90° bend to achieve less energy loss. In addition to changing the direction of EM wave propagation, the structure can also further improve the purity of linear polarization of the system by filtering the vertical EM waves.

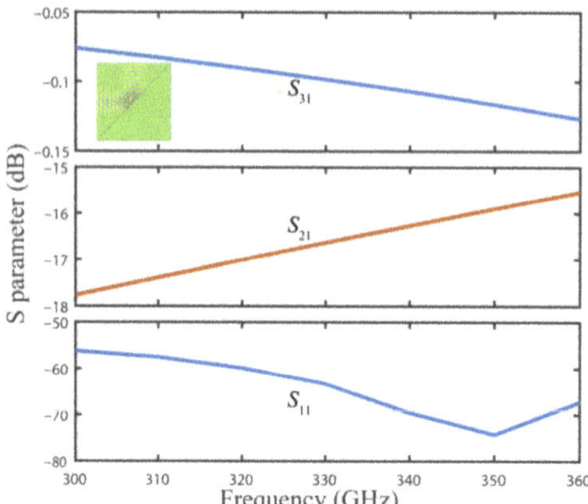

Figure 16. Simulation results of 90° bend waveguide in the frequency band of 300–360 GHz.

4. Conclusions

In this paper, a TE_{11}-HE_{11} mode converter and a three-port directional coupler for the 330 GHz/500 MHz DNP-NMR system transmission line are proposed. The optimized converter can maintain good linear polarization performance, which is 98.8% at 330 GHz. Besides, it can produce a pure HE_{11} mode to facilitate subsequent low loss transmission. The three-port directional coupler is designed for monitoring the EM wave power in the transmission line in real-time, and it can stably output approximately −33 dB of the electromagnetic wave power on port 3 at the frequency range of 300−360 GHz. Directivity can reach up to 11.8 dB at 330 GHz. The designed mode converter and three-port directional coupler can be easily installed in the DNP-NMR transmission system.

Author Contributions: Conceptualization, X.Y., P.L. and C.D.; Formal analysis, X.Y. and C.D.; Funding acquisition, P.L. and C.D.; Investigation, C.D.; Methodology, X.Y. and C.D.; Software, X.Y.; Supervision, P.L. and C.D.; Validation, Z.Z.; Writing—original draft, X.Y.; Writing—review and editing, C.D., J.Z. and T.H. All authors have read and agreed to the published version of the manuscript.

Funding: This research was funded by National Natural Science Foundation of China under grant number 61531002, 61861130367, NSAF-U1830201. It was also supported in part by the Newton Advanced Fellowship from Royal Society (NAF/R1/180121), United Kingdom.

Data Availability Statement: All data included in this study are available upon request by contactingwith the corresponding author.

Acknowledgments: The authors thank Bao-Liang Hao from Beijing Vacuum Electronics Research Institute, China, for his support for CST software and professional suggestions in the paper.

Conflicts of Interest: The authors declare no conflict of interest.

References

1. Nanni, E.A.; Barnes, A.B.; Griffin, R.G.; Temkin, R.J. THz Dynamic Nuclear Polarization NMR. *IEEE Trans. Terahertz Sci. Technol.* **2011**, *1*, 145–163. [CrossRef] [PubMed]
2. Bayro, M.; Debelouchina, G.T.; Eddy, M.T.; Birkett, N.R.; MacPhee, C.; Rosay, M.; Maas, W.E.; Dobson, C.M.; Griffin, R.G. Intermolecular structure determination of amyloid fibrils with magic-angle spinning and dynamic nuclear polarization NMR. *J. Am. Chem. Soc.* **2011**, *133*, 13967–13974. [CrossRef] [PubMed]
3. Rossini, A.; Zagdoun, A.; Lelli, M.; Lesage, A.; Copéret, C.; Emsley, L. Dynamic nuclear polarization surface enhanced NMR spectroscopy. *Acc. Chem. Res.* **2013**, *46*, 1942–1951. [CrossRef]
4. Abragam, A.; Goldman, M. Principles of dynamic nuclear polarization. *Rep. Prog. Phys.* **1978**, *41*, 395. [CrossRef]
5. Kemp, T.F.; Dannatt, H.R.W.; Barrow, N.S.; Watts, A.; Brown, S.P.; Newton, M.E.; Dupree, R. Dynamic nuclear polarization enhanced NMR at 187 GHz/284 MHz using an extended interaction Klystron amplifier. *J. Magn. Reson.* **2016**, *265*, 77–82. [CrossRef] [PubMed]

6. Blank, M.; Felch, K.L. Millimeter-Wave Sources for DNP-NMR. *Emagres* **2007**, *7*, 155–166.
7. Mendis, R.; Grischkowsky, D. Plastic ribbon THz waveguides. *J. Appl. Phys.* **2000**, *88*, 4449–4451. [CrossRef]
8. Jamison, S.P.; McGowan, R.W.; Grischkowsky, D. Single-mode waveguide propagation and reshaping of sub-ps terahertz pulses in sapphire fibers. *Appl. Phys. Lett.* **2000**, *76*, 1987–1989. [CrossRef]
9. De Rijk, E.; Macor, A.; Hogge, J.-P.; Alberti, S.; Ansermet, J.-P. Note: Stacked rings for terahertz wave-guiding. *Rev. Sci. Instrum.* **2011**, *82*, 066102. [CrossRef]
10. Wang, K.; Mittleman, D.M. Metal wires for terahertz wave guiding. *Nature* **2004**, *432*, 376–379. [CrossRef]
11. Bowden, B.; Harrington, J.A.; Mitrofanov, O. Low-loss modes in hollow metallic terahertz waveguides with dielectric coatings. *Appl. Phys. Lett.* **2008**, *93*, 181104. [CrossRef]
12. Pike, K.J.; Kemp, T.F.; Takashi, H.; Day, R.; Howes, A.P.; Kryukov, E.V.; MacDonald, J.F.; Collis, A.E.C.; Bolton, D.R.; Wylde, R.J.; et al. A spectrometer designed for 6.7 and 14.1 T DNP-enhanced solid-state MAS NMR using quasi-optical microwave transmission. *J. Magn. Reson.* **2012**, *215*, 1–9. [CrossRef]
13. Zhang, L.; He, W.; Donaldson, C.R.; Smith, G.; Robertson, D.A.; Hunter, R.I.; Cross, A.W. Optimization and measurement of a smoothly profiled horn for a W-band gyro-TWA. *IEEE Trans. Electron. Devices* **2017**, *64*, 2665–2669. [CrossRef]
14. Imai, T.; Kobayashi, N.; Temkin, R.; Thumm, M.; Tran, M.; Alikaev, V. ITER R&D: Auxiliary systems: Electron cyclotron heating and current drive system. *Fusion Eng. Des.* **2001**, *55*, 281–289.
15. Teniente, J.; Gonzalo, R.; Rio, C.D. Low sidelobe corrugated horn antennas for radio telescopes to maximize G/Ts. *IEEE Trans. Antennas Propag.* **2011**, *59*, 1886–1893. [CrossRef]
16. Addamo, G.; Peverini, O.A.; Tascone, R.; Virone, G.; Cecchini, P.; Orta, R. A Ku-K dual band compact circular corrugated horn for satellite communications. *IEEE Antennas Wirel. Propag. Lett.* **2009**, *8*, 1418–1421. [CrossRef]
17. Simonetto, A.; Solari, G.; Gandini, F.; Granucci, G.; Muzzini, V.; Sozzi, C. Directional couplers-polarimeters for high-power corrugated waveguide transmission lines. *Fusion Sci. Technol.* **2001**, *44*, 247–252. [CrossRef]
18. Woskov, P.P.; Bajaj, V.S.; Hornstein, M.K.; Temkin, R.J.; Griffin, R.G. Corrugated waveguide and directional coupler for CW 250-GHz gyrotron DNP experiments. *IEEE Trans. Microw. Theory Tech.* **2005**, *53*, 1863–1869. [CrossRef]
19. Granet, C.; James, G.L. Design of corrugated horns: A primer. *IEEE Antennas Propag. Mag.* **2005**, *47*, 76–84. [CrossRef]
20. Clarricoats, P.J.B.; Olver, A.D. *Corrugated Horns for Microwave Antennas*; Peter Peregrinus Ltd.: London, UK, 1984.
21. Nanni, E.A.; Jawla, S.K.; Shapiro, M.A.; Woskov, P.P.; Temkin, R.J. Low-loss transmission lines for high-power terahertz radiation. *J. Infrared Millim. Terahertz Waves* **2012**, *33*, 695–714. [CrossRef]
22. Wylde, R.J. Millimetre-wave Gaussian beam-mode optics and corrugated feed horns. *IEE Proc. H Microw. Opt. Antennas* **1984**, *131*, 258–262. [CrossRef]
23. Roberts, D.A.; Rahm, M.; Pendry, J.B.; Smith, D.R. Transformation-optical design of sharp waveguide bends and corners. *Appl. Phys. Lett.* **2008**, *93*, 251111. [CrossRef]
24. Tang, H.H.; Huang, B.; Huang, T.J.; Tan, Y.; Liu, P.K. Efficient waveguide mode conversions by spoof surface plasmon polaritons at terahertz frequencies. *IEEE Photonics J.* **2017**, *9*, 1–10. [CrossRef]
25. Thumm, M.K.; Kasparek, W. Passive high-power microwave components. *IEEE Trans. Plasma Sci.* **2002**, *30*, 755–786. [CrossRef]

MDPI
St. Alban-Anlage 66
4052 Basel
Switzerland
Tel. +41 61 683 77 34
Fax +41 61 302 89 18
www.mdpi.com

Electronics Editorial Office
E-mail: electronics@mdpi.com
www.mdpi.com/journal/electronics

www.ingramcontent.com/pod-product-compliance
Lightning Source LLC
LaVergne TN
LVHW070905170726
843515LV00005B/706

9783036554488